Kathleen S. Ziprik

PR Power

Public Relations for Building Pros

BuilderBooks™
National Association of Home Builders
1201 15th Street, NW
Washington, DC 20005-2800
www.builderbooks.com

PR Power: Public Relations for Building Pros
Kathleen S. Ziprik

ISBN 0-86718-545-7

© 2003 by BuilderBooks™
of the National Association of Home Builders
of the United States of America

All rights reserved. No part of this book may be reproduced or utilized in any form or by any means, electronic or mechanical, including photocopying and recording or by any information storage and retrieval system without permission in writing from the publisher.

Cover design by Armen Kojoyian
Printed in the United States of America

Cataloging-in-Publication Data

Ziprik, Kathleen S., 1959–
   PR Power : public relations for home builders / Kathy S. Ziprik.
     p. c.m.
Includes bibliographical references and index.
   ISBN 0-86718-545-7 (pbk.)
1. Construction industry–Public relations.   2. Construction industry–Customer services.
3. Building trades–Public relations.   4. Building trades–Customer services.
5. Contractors–Public relations.   6. Contractors–Customer services.   I. Title.

HD9715.A2 Z56 2002
659 .2'96908–dc21

2002151501

Disclaimer
This publication is designed to provide accurate and authoritative information in regard to the subject matter covered. It is sold with the understanding that the publisher is not engaged in rendering legal, accounting, or other professional service. If legal advice or other expert assistance is required, the services of a competent professional person should be sought.
–From a Declaration of Principles jointly adopted by a Committee of the American Bar Association and a Committee of Publishers and Associations.

For further information, please contact:
BuilderBooks™
National Association of Home Builders
1201 15th Street, NW
Washington, DC 20005-2800
(800) 223-2665
Check us out online at: www.builderbooks.com

12/02 SLR/ Data Reproductions Corp., 1500
Public Relations for Home Builders Book

About the Author

Kathleen Schoch Ziprik has been practicing public relations for more than 20 years in the building product and hospitality industries. Since 1996 she has operated her own public relations firm based in Douglasville, Georgia, and focuses on assisting building product companies with their public relations needs. Her current and past clients include Style Solutions Inc., Simonton Windows, Hy-Lite Products Inc., Metallon Inc., ODL Inc., Woodcraft Supply Corp., Owens Corning, Weather Shield Windows & Doors, Isobord Enterprises, Vengeance Creek Stone, Stephen Fuller Inc., and Hinge-It Corporation.

Prior to starting her own firm, Ziprik launched her interest in the building industry by serving as public relations manager for Georgia-Pacific Corp. in the building products division for several years. Other previous experiences include public relations and marketing positions with Hyatt Hotels and Six Flags Corporation.

A graduate of Rowan University and the University of Georgia with B.A. and M.A. degrees in public relations, Ziprik has a long history of involvement with the Public Relations Society of America (PRSA) and the Public Relations Student Society of America (PRSSA). Starting in college, she served as the National Chairman of the 4,300-member student organization. Since then she has taught public relations courses at Mercer University and served as co-chairman of the Champions of PRSSA along with leadership roles in several PRSA sections. She was the first inductee into the PRSSA National Hall of Fame in 1993 and received the First Annual Distinguished PR Alumni Award from Rowan University in 1988.

This is Ziprik's second time authoring a book for BuilderBooks. Her first book, *PR Power: Public Relations for Building Products Manufacturers,* was published in 2002.

Acknowledgments

Many thanks to all the individuals who assisted with information, quotes, and materials for this book. A wide variety of people within the industry shared their insights so that this book would be especially valuable for home builders of all sizes and in different geographic locations.

In particular, there are several people who have been a pleasure to work with on almost a daily basis over the years. While some would call them clients, to me they're a great deal more. These are exceptionally talented individuals in their own areas of expertise who truly understand and appreciate the value of public relations: John Brunett, Charlie Crooks, Jean Dimeo, Shawn Draper, Joy Frank-Collins, Gary Good, Dave Goulette, Karl Hatrak, Bryan Katchur, Angelo Marasco, Tina Mealer, Chris Monroe, Mike Reed, Sam Ross, Steve Roth, Mike Shon, and Craig Weaver.

Warm thanks to Theresa Minch, Jessica Poppe, Stacy Stufft, and Elizabeth Christy for reviewing the book and assisting with valuable comments.

For "support on the home front" and daily encouragement with writing this book and all my projects, thanks to: Joyce and Fred Schoch; Genie and Tod Gordon; Kim Drew, APR; Nancy and Don Zeman; Barbara, Mike, and Matthew Robeson; and always, to Brandy Alexander Ziprik.

Finally, every person should have a mentor and champion in their corner. I've always had the best–Anthony J. Fulginiti, APR–public relations practitioner extraordinaire, teacher, and friend.

Book Production

PR Power: Public Relations for Building Pros was produced under the general direction of Gerald Howard, NAHB Executive Vice President and CEO, in association with NAHB staff members Michael Shibley, Executive Vice President, Builder, Associate, and Affiliate Services; Greg French, Staff Vice President, Publications and Non-dues Revenues; Eric Johnson, Publisher, BuilderBooks; Theresa Minch, Executive Editor; and Jessica Poppe, Assistant Editor.

Contents

Chapter 1: Unraveling the Mysteries of Public Relations — 1
  Defining Public Relations for Home Builders — 1
  Public Relations: An Important Piece of the Marketing Pie — 2
  Maximizing Marketing Investments with Public Relations — 7
  Who Should Use Public Relations? — 7
  The Finish Line: What to Expect from Public Relations — 8

Chapter 2: Getting Started — 11
  Assigning the Public Relations Function — 11
    On-Staff Public Relations Experts — 12
    On-Staff Communications/Marketing People — 12
    Public Relations Interns — 13
    Outsourcing to Public Relations Agencies — 14
    Independent Public Relations Consultants — 16
    Prioritizing Public Relations Goals for Your Company — 17
  Identifying Your Target Audiences — 18
  Plan the Work and Work the Plan — 19
  Public Relations Tools — 21
    Press Releases and Press Kits — 21
    Visual Support: Slides, Photos, CDs, Videos, Transparencies, and Websites — 25
    By-Lined Stories — 27
    Industry Trends Stories — 29
    Media Binders — 30
    Deskside Briefings — 31
    Subdivision and/or Show Home Tours — 33
    Media Placement Services — 34
    Company Literature — 34
  Budgeting for Public Relations — 35
  Tracking for Success — 37
    Measurement Tools and Resources — 37
    Sharing the Good News — 38

Chapter 3: Tackling Special Events — 40
Defining Special Events for Home Builders — 40
Logistics of Special Events — 40
Timeline for Success — 44
Checklist of Winning Special Event Ideas — 44
Maximizing Special Events — 46

Chapter 4: Spotlight on Media Relations — 47
Understanding the Media — 47
The Five Ws of Media Relations — 47
   Types of Media and the Unique Deadlines — 50
   No Promises — 57
   Killing the Myth: Advertising Does NOT Guarantee Story Placement — 58
Dealing with the Media — 58
   Care and Feeding of the Media — 58
   What to Expect From the Media — 59
   What the Media Expect From You — 60
Launching Into Media Relations — 61
   Identify the Right Media for Your Message — 62
   Planning for Media Attention — 63
   Assigning a Spokesperson — 65
   Do's and Don't's of Working with the Media — 67
   "Faster-Than-A-Speeding-Bullet" Responses — 70
Making Your Story Newsworthy — 70
   Taking Advantage of Current Trends — 72
   Leads On Unique Projects — 75
Positioning Your Company as Experts: How to Get Exposure When There Is No Hard News — 76

Chapter 5: Winning At Community Relations — 79
The Perfect Fit: Getting Involved in Projects and Programs that Make Sense for Your Company — 79
Charitable Giving Versus Volunteerism: Finding the Perfect Balance for Your Company — 80
Checklist of Local Opportunities — 82
Leading the Way: Creating a Solid Community Image — 84

The Importance of Local Home Shows	85
Maximizing Community Relations Efforts	87
Sharing the Good News	88

Chapter 6: Partnering for Success — **90**

Hitch a Ride–Working Through and With Local Home Builder Associations	90
Joint Builder Efforts	92
Promoting With Manufacturers	94
Working with Dealers	96
Special Magazine and TV Show Projects	99
Getting Former Customers to Promote Your Company	102

Chapter 7: Employee Relations — **106**

Sharing the Vision and the Message	106
Getting Your Team to Promote the Company	107
Tools of Employee Relations	108
Sharing Positive Results	110

Chapter 8: Promotions and Publicity — **112**

Ideas That Set Your Business Apart	112
Rule #1: Maximize Exposure	114

Chapter 9: Planning for a Crisis — **117**

It CAN Happen to You: A Look at Scenarios Facing Home Builders	117
You've Got to Have a Plan	119
Preparing for the Unexpected	119
Sharing the Plan With Others	120
Designating a Spokesperson	121
Practice, Practice, Practice	122
Do's and Don'ts of Media Interviews During a Crisis Situation	125

Appendix: Resources — **127**

Index — **129**

Figures

Chapter 1: Unraveling the Mysteries of Public Relations   1

Figure 1-1. Bylined articles in consumer and trade magazines, like the one shown here from *Architectural West*, can be an important piece of the overall marketing pie.   4

Figure 1-2. Whether you build tract homes, custom homes, or multi-housing living spaces, all builders should use public relations tactics to enhance their business.   5

Figure 1-3. A variety of companies are available to help builders target their messages to key audiences and then track results through clipping services.   5

Chapter 2: Getting Started   11

Figure 2-1. Public relations interns can help builders handle special events, track press clippings and put together press kits.   14

Figure 2-2. Special companies, such as Burrelle's Information Services, exist to help their customers identify and reach key target audiences.   19

Figure 2-3. Builders should follow the same format as manufacturers and other companies when issuing press releases–all the critical 5 Ws go in the first paragraph.   22

Figure 2-4. CDs of information and photos, like this one prepared by St. Thomas Creations, can quickly educate media people on your company and the support materials you have available.   26

Figure 2-5. This bylined story in *Architectural West* magazine brings tremendous exposure to Simonton Windows with architects–one of their key target audiences.   29

Figure 2-6. Industry trends stories–whether on product advancements or changes in consumer buying habits–are always popular with industry publications.   30

Figure 2-7. Media binders are an excellent way to present all your company information and public relations materials in one comprehensive unit to media people.   31

Figure 2-8. Remember to maximize your company literature by sharing brochures, sell sheets and flyers with your key target media members.   35

Figure 2-9. Several companies offer press clipping services. Using one of these companies assures that you don't miss any media coverage your company might generate.   38

Figure 2-10. Maximize your press clips by posting them on employee bulletin boards, sharing them with your sales team and creating a "brag book" for your reception area.   39

Chapter 3: Tackling Special Events   40

Figure 3-1. At this Owens Corning System Thinking Show Home, the builder in Ohio was all smiles–a grand opening special event for the home brought more than 3,000 people through its doors in one weekend.   41

Figure 3-2. Special events require lots of details . . . right down to the flowers, goodie bags and bagels at this open house on a trade show floor.   43

Figure 3-3. Invitations to special events can take many forms, including these simple flyers. Remember to send out invitations at least 4–5 weeks in advance for a major event.   43

Chapter 4: Spotlight on Media Relations   47

Figure 4-1. Contractor Bruce Rosenthal understands the media. That's why he wore his "Handyman Life Member" jacket to an installation photo shoot for miterless corners for Style Solutions. His quick thinking of wearing the jacket allowed the manufacturer to gain a four-page story in *Handy* magazine on Bruce's installation project.   48

Figure 4-2. The Burrelle's directories can easily help a builder correctly identify the specific media outlets he needs for his press releases and materials.   51

Figure 4-3. Consumer magazines can be both general in nature and themed to home improvement. Don't overlook any magazine as a potential showcase for gaining publicity for your homes.   52

Figure 4-4. Media directories, like this one on newspaper listings from Burrelle's, are an excellent way to develop a comprehensive and targeted media list for your business.   54

Figure 4-5. Most newspaper and magazine editors adhere to writing standards set forth in the *Associated Press Stylebook and Libel Manual*. Using this manual for your own writing will help your releases be more readily acceptable by the media.   57

Figure 4-6. "Hot" topics for the media always include low maintenance products. If your company is using innovative products, such as the urethane millwork shown here being installed on a Habitat project, offer to serve as a spokesperson to the local media.   66

Figure 4-7. When contractors for Heivilin Remodeling took on a renovation project in Virginia, they quickly found themselves the focal point of stories in *Professional Remodeler* magazine. The unique project brought the company both print and broadcast media attention.   71

Figure 4-8. Have you recently built a home with lots of windows and doors that allow light and air to flow through a home? If so, consider creating a story for your local media on the connection homeowners are seeking between their interior and exterior living space. Any home has an angle for a story . . . it's the job of the public relations professional to find that angle and maximize it. (Courtesy of Crossville Porcelain Stone/USA).   73

Figures **xi**

Chapter 5: Winning At Community Relations 79
Figure 5-1. On a job site at the 2002 International Builders' Show, builders and manufacturer's representatives volunteered to construct five homes for Habitat for Humanity that were later moved to different locations in Atlanta. 81
Figure 5-2. Whether your company volunteers manpower for a national or local project, the important factor is to link in with projects that support the housing industry or relate to construction in some way. 83
Figure 5-3. Contact your local chamber of commerce, United Way or American Red Cross chapters to determine what construction projects in your area can benefit best from your company's involvement. Before committing to a project, ask to speak with the organization's public relations person to determine how your efforts will be promoted in the marketplace. 84

Chapter 6: Partnering For Success 90
Figure 6-1. Looking for a unique way to gain publicity? Create a collection of playhouses and then raffle them off for charity at a local mall or state fair. (Courtesy of Style Solutions). 95
Figure 6-2. This home, built by Dobson Construction, was featured as a cover story in the August 2002 issue of *Luxury Home Builder*. Public relations people at Simonton Windows® worked to "place" the story in the magazine–and also created a four-page case study brochure on the home. 98
Figure 6-3. Many manufacturers create case study brochures showcasing unique projects that use their products. Connecting with manufacturers on your homes can result in free publicity and sales pieces for your company. (Courtesy of Simonton Windows®) 98
Figure 6-4. Home improvement television shows, like the ones shown here, are always eager to learn about unique home remodeling and new construction projects. Having professional photography taken of your home to "pitch" to television show producers is the first step in getting on the air. 100
Figure 6-5. Building product manufacturers, like Weather Shield Windows and Doors, oftentimes include builder projects in their newsletters, customer communications materials and website. 101
Figure 6-6. Get it in writing. If you've created the perfect home for a customer, don't hesitate to request a letter from them singing your praises. Also, offer to send a photographer to the home once they're moved in, so that you both have professional images of your showcase home to share with future potential customers. (Photo courtesy of Style Solutions) 103

Chapter 7: Employee Relations 106
Figure 7-1. Make certain that every management team member, employee and subcontractor know and understand the vision for your company.

Only when everyone knows of your deep commitment to your business can they embrace the same level of dedication. (Photo courtesy of Style Solutions)   107

Figure 7-2. Wondering how to reach your construction team with your company messages? Try paycheck stuffers, direct mailed literature to their homes, employee newsletters and worker incentive programs. (Photo courtesy of Style Solutions)   109

Figure 7-3. Share publicity results with your employees by posting clips on bulletin boards, including them with paychecks, or, as shown here, creating special CDs with publicity results for them to view at home.   110

Chapter 8: Promotions and Publicity   112

Figure 8-1. Are you installing high-end kitchen appliances in your home? If so, contact the manufacturer. See if they can work with you on creating a cooking demonstration in your market, exhibit at a local home show or even create a cookbook for your customers. "Push the envelope" to set your business apart in the marketplace.   113

Chapter 1

Unraveling the Mysteries of Public Relations

Welcome to the wonderful world of public relations. Want a simple definition for public relations? Think of it as the process of trying to influence people.

Basically, builders can use public relations strategies to influence people to have a specific opinion about their company and their homes–and that opinion can influence a consumer towards a home purchase. These strategies include issuing press materials to local media, creating sales tools, giving speeches to community organizations, or even developing a website. Many of these activities may already be underway at your company.

Regardless the size of your company, public relations should be an integral part of your marketing operations. However, just as you can't build a house in three weeks, neither can you launch a successful public relations campaign in a short time. Plan for the long term and your program will be beneficial.

Defining Public Relations for Home Builders

There is no mystery to public relations, but there certainly are some misconceptions. How many times have you heard someone say, "Oh, you're so good with people, you ought to be in public relations"? The truth is, you can be terrible with people and still possess outstanding capabilities as a public relations professional. Or, you can be wonderful in dealing with people yet have no communication or organizational talents–two of the most critical elements of public relations.

Public relations is a function of communication. No matter what company, organization, or association employs a public relations campaign, the purpose usually is for communicating messages. Sometimes the focus is on internal audiences (such as employees and shareholders); other times, the focus is to communicate externally, with audiences such as customers, the general public, or the media.

Although public relations is a general "catch-all" name for the profession, public relations encompasses many subcategories and elements. Some areas are covered in this book because they apply to home builders. However, the scope of public relations entails so many areas that it would be difficult to contain them all within these pages.

A public relations practitioner can be a generalist, handling all realms of public relations, or a specialist, handling any (or a combination) of the following areas:

- Financial relations
- Community relations
- Publicity
- Special events
- Employee relations
- Speechwriting
- Media relations
- Public affairs
- Research
- Government affairs and lobbying
- Issues management
- Fundraising

> Many different elements can help set the stage for home sales. Public relations is just one ingredient in your company's recipe for success.

> Want to gain a deeper understanding of public relations? Are you searching for a public relations professional to join your team? Then contact the Public Relations Society of America at www.prsa.org. They have 20,000 members nationwide, a professional job hotline, and lots of resources to help you find a public relations practitioner on either a temporary or a long-term basis.

Public Relations: An Important Piece of the Marketing Pie

Think of public relations as part of the marketing pie. Within the scope of a marketing program, builders of almost every size use marketing tools such as advertising, direct mail, literature and sales aids, and websites to develop their brands. All of these pieces "fit together" to achieve a company's overall marketing and sales goals. And all of these pieces help build credibility for a company and support overall marketing goals.

Public relations should be part of your marketing mix, or marketing pie. Why? Because it makes good strategic business sense and because your public relations efforts can easily complement the other marketing elements your company undertakes.

Here's an example: When a company like Valentini Builders constructs, furnishes, and staffs a show home in Walnutta, Colorado, they create and implement a strategic marketing campaign. Part of the introductory program may be to advertise in targeted consumer newspapers and magazines. A brochure may be created on the home as a hand-out. Invitations may be mailed to potential home buyers to come visit the show home. And the builder's website may include a big, splashy story on the show home.

Where does public relations come in? Go

> "Builders and companies that have never before used public relations strategies are amazed at how valuable this investment can be. Once their eyes are opened to the benefits of incorporating public relations in their marketing mix, they never go back to pre-PR times. They've found the investment is far too valuable for them to ignore."
>
> **Margie Simon**
> **Principal**
> **Simon & Associates Public Relations**

back to the magazines and newspapers targeted to consumers. Press releases and photos can be sent to those publications introducing the show home, resulting in "free" publicity in the magazines. Why is that important if your company advertises in publications? For several reasons.

First, consider it a "one-two punch." If a person sees an ad for the open house *and* reads about it in a newspaper or magazine, that's two impressions versus one on the target audience.

Second, how many people do you know who buy newspapers and magazines simply to read the ads? None. People select and buy their publications for the editorial content. Ads are often perceived as a "necessary evil" that readers "put up with" in order to read the contents. The advertisements are secondary. So it benefits Valentini Builders greatly in this fictitious case to have editorial coverage of its show home in the magazine because editorial content is why people read a magazine.

Third, people expect an advertisement to be one-sided. Valentini Builders isn't going to take out an ad showcasing any negative aspects of their new show home—ads are self-serving—and the general public recognizes this. Press coverage is not one-sided. So when the editors of the *Walnutta News* run a story with photos on the exciting features of the new show home, it's akin to the newspaper saying, "Hey, this home looks pretty good to us—we encourage you to visit it and consider buying." In a sense, the publicity placement becomes a type of third-party endorsement for Valentini Builders.

> "Public relations professionals know how to get reporters' attention, suggest stories, and provide the materials journalists need. These are skills most builders don't have (just as most PR people couldn't build a house). Builders are accustomed to hiring subcontractors to perform specific tasks to get a house built. Think of a public relations person as another 'subcontractor' who helps builders accomplish another task: getting the house sold."
>
> Judy Stark
> Home Editor
> *St. Petersburg Times*

> "Opportunities for publicity through newspapers are endless. You may get coverage at a local newspaper for your company's involvement in a Habitat For Humanity project, volunteering to help clean a park, the grand opening of a show home, and/or serving as an expert on housing trends issues. The key to success is making local editors aware of your capabilities on a regular basis so they know who to call when they need information for stories."
>
> Julie Chalpan
> President
> Focus Unlimited

Figure 1-1. Bylined articles in consumer and trade magazines, like the one shown here from *Architectural West,* can be an important piece of the overall marketing pie.

Fourth, in many cases public relations is more affordable than placing an advertisement. While there are no guarantees that a press release, a story, or product review will run (as opposed to an ad, in which you can specifically target when and where it will get placed), some companies do not have abundant budgets for advertising. Public relations efforts are highly cost effective because you're paying to send information and photography to publications, which will hopefully run the materials at no cost to you.

Does this mean your company should eliminate advertising and rely solely on public relations efforts? No. Public relations should be a complementary part of the marketing mix, a well-balanced mix. If your company has a limited advertising budget, you may be able to target several important publications in your market for advertising and then supplement marketing efforts with press releases and materials to dozens, or hundreds, of other publications. And, don't forget, you can send these same media materials to radio shows and television shows to hopefully enhance awareness of your homes.

Another option to consider is strategic advertising. Some builders advertise only to trade audiences (such as potential move-up buyers in distinct cities) because of limited budgets. But these builders enhance their advertising buys by sending out press releases to a variety of media outlets (including magazines, newspapers, radio shows, and TV shows) to generate a groundswell of media attention in markets they can't afford to advertise in.

Think back to the idea that Valentini Builders is introducing a show home.

Unraveling the Mysteries of Public Relations 5

Figure 1-2. Whether you build tract homes, custom homes, or multi-housing living spaces, all builders should use public relations tactics to enhance their business.

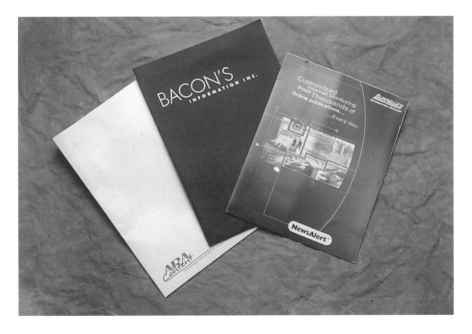

Figure 1-3. A variety of companies are available to help builders target their messages to key audiences and then track results through clipping services.

After the public relations function, budget, and goals have been established and assigned, it's time to launch a game plan. Although this entire book is meant to serve as a guideline for creating a comprehensive plan, some builders just starting a public relations program may benefit from the following checklist of action steps.

- Identify target audiences for your company messages and create media lists to complement those target audiences.
- Research and create a comprehensive written public relations plan that reflects the company's overall objectives and goals.
- Evaluate all existing company photography to determine what can be used for media purposes. Create duplicate slide images and/or CDs for media use of existing images. Schedule photo shoots for outstanding projects needing new photography.
- Evaluate existing company literature and determine which pieces would help media members better understand your company.
- Assign and train a company spokesperson.
- Research and develop a comprehensive company press kit, including a backgrounder release on the company, a company fact sheet, an overview of your homes, and any pertinent company literature.
- Determine which trade shows the company will participate in for the year. Develop pre-show, show, and post-show public relations strategies.
- Support major show home and new home/development openings with press releases and images distributed to the appropriate target media to maximize company messages.
- Determine if the public relations function can support community and employee

Public relations can help by coordinating tours of the home from the construction stage through completion to key media members in their area. A step-by-step series of articles could be arranged with the paper to follow the entire construction process so that home buyers gain a better understanding of the building process.

In the case of publicizing the new home on the company's website, talented public relations people can supply copy and direction for the introduction of the show home on the site. With direct mail programs, again public relations people can help in the creation and writing of the pieces along with the overall direction for external communications. You may even be able to use the public relations team to help generate internal enthusiasm for the new show home through employee announcements, newsletters, and meetings.

Ideas for incorporating public relations into a home builder's overall marketing program are endless. This book includes hundreds of ideas–customized for the builder–to help put the power of public relations to work at your company.

Maximizing Marketing Investments with Public Relations

Eager to make your marketing dollars stretch as far as they can? Then you definitely should consider using public relations strategies.

When you invest in advertisements and direct mail programs to let people know about your homes, you're marketing to them. You can enhance those marketing efforts by using public relations strategies.

Does your company target first-time home buyers? If so, consider hosting regular "Tips and Strategies for First-Time Home Buyers" sessions at your offices, model home, local schools, realtor offices, or community gathering locations. Why? Because you want to educate and form a bond with people who are getting ready to purchase a home. Then, when they see your ads and receive your direct mail pieces, there will be an immediate connection.

Not interested in first-time home buyers but looking for baby boomers planning for retirement? Link up your efforts with a financial organization to offer sessions and ideas on saving for and buying the home of their dreams. Create a series of stories on the same topic and offer it to your local newspaper. Present the idea to a local television and/or radio station and get more free publicity. The bottom line is that public relations efforts can help strengthen and build on your existing marketing programs.

- relations programs in place or if they need to be created and implemented.
- Research, write, and practice a full crisis communications program involving all key management team members.
- Start pitching softer news stories, such as industry trends, to media representatives to maximize company exposure throughout the year.
- Set up evaluation methods for public relations activities, including clipping services, recording inquiry calls and website hits, administering questionnaires, etc.

In the marketing arena, public relations should never be treated like a stepchild. The public relations function should go hand-in-hand with all other marketing and sales programs to help a company reach its sales goals.

"As a small-volume custom builder, I handle the public relations aspects myself for our company. We've found value in incorporating public relations strategies into our overall business plan as a way to strengthen our presence in the community and build business."

Rick Buttorff
Owner
The Buttorff Company

Who Should Use Public Relations?

Think public relations efforts should be reserved for large builders only? Think again. Public relations is a powerful tool that builders of all sizes can implement.

A custom builder who constructs just four homes a year can benefit from solid public relations tactics just as much as a tract home builder creating several development sites each year. The key is not how big (or small) a builder you are but how effectively you create and implement public relations strategies that will bring your

> Although it's entirely possible to turn public relations programs on and off like tap water, it's not advisable. Solid, long-term public relations plans and campaigns help steadily build the reputation and image of your company. Well thought-out public relations programs can have lasting and far-reaching benefits for builders.

company recognition. Remember: recognition equals awareness. People must be aware of you and your homes before they can purchase them.

The Finish Line: What to Expect from Public Relations

There's no doubt about it—savvy builders incorporate elements of public relations into both their company business and their marketing programs. However, measuring the results of these efforts can be challenging.

One of the toughest aspects of any public relations campaign can be trying to quantifiably measure its results. How do you put a dollars and cents figure to what your company gets out of donating labor to a Habitat For Humanity project? It's impossible to put a price tag on the goodwill and community spirit you can achieve with such an important contribution of labor to a community project, yet everyone at your company will generally agree that the experience and involvement were valuable.

> Companies with smaller advertising budgets can turn to public relations efforts to maximize exposure in target markets. The combined effort of limited ads and multiple public relations programs can make for a powerful presence in the community. Even when a company grows and can afford to spend more dollars on advertising, it's wise to retain and strengthen public relations efforts. Builders should not view advertising and public relations as an "either/or" choice—rather, as two complementary marketing elements that help drive success for their companies.

Oftentimes, building product manufacturers face the same question of measuring results as do builders when faced with public relations programs. Consider Georgia-Pacific Corporation. How does the industry giant actually know how many sheets of plywood a builder buys each year just because the company has donated stacks of the product to the television show *This Old House?* It's virtually impossible. Even though large blue Georgia-Pacific logos show up on much of the lumber used on the popular television show, there's no way to figure out how much additional product is sold because of the association the company has achieved with the leading home improvement show.

In this case, Georgia-Pacific is relying on the product donation as just one element of its public relations campaign. If you, as a builder, see the product used as a preferred product on the show, you may think twice the next time you

> The more a potential home buyer sees your company name in a positive way, the more memorable their impression. So use public relations strategies and tactics to strengthen your company image.

purchase plywood. And the folks at Georgia-Pacific are hoping you'll remember their involvement in *This Old House*. Their goal is to get you to purchase Georgia-Pacific plywood. So whether you see an ad in a magazine for the product or the plywood used on the popular television show, they're trying to "impact" you with their product messages as much as possible.

Should measurement of public relations programs be considered a lost cause? No. First of all, consider the intrinsic value of supporting your other marketing and sales programs with public relations campaigns and plans. Immediately your company is strengthening its message reach to target audiences.

On a more tangible level, some public relations efforts can indeed be quantitatively measured. Press clips offer a valid and solid look at the success of story placements, press release distribution, and media coverage for a builder.

Several companies such as Luce Press Clippings, Burrell's Information Services, and Bacon's Information, Inc. offer the service of researching, clipping, and sending company stories to you that mention the company name, products, or other key "search words" you select. There is generally a flat monthly search fee plus a "per clip" fee for this service.

Depending on how much press coverage your company receives, a clipping service can cost you several hundred or several thousand dollars a year. Most public relations professionals treat this as a wise investment.

Without a clipping service, builders have no true understanding of the successful placements they achieve with their public relations efforts. If your company is isolated to just one geographic town or county, it may be possible to monitor the media yourself for press coverage. However, most builders have homes for sale in dozens of counties, or even states. Those builders can't monitor the media every day in all those locations. It's simply unrealistic to think that you will "find" your placements by monitoring newspapers and magazines yourself. A clipping service offers a fast and effective way to locate and measure successful placements.

> "Our marketing and public relations efforts really skyrocketed when our company was selected to construct the Design Home 2002 in conjunction with *Philadelphia* magazine. Not only was our company promoting the show home, but the magazine and charity involved promoted it—as did dozens of sponsors of the home.
>
> Even after the show home closed, we were still getting calls from manufacturers whose products were in the house. They were interested in doing public relations on their products and the home—which naturally included us. This one project alone really showed me how public relations can boost the reputation and awareness of our company in the marketplace."
>
> Matt Thompson
> President
> Thompson Homes, Inc.

> "By donating lumber products to the best home improvement television shows, we create awareness for our products. From there, our marketing efforts focus on building familiarity and preference for Georgia-Pacific products. After that, we hope to make a sale. The public relations value we get from aligning ourselves with shows like *This Old House* is immeasurable. The partnership with the show helps us create an image that sells products."
>
> Francis Giknis
> Director of Corporate Marketing
> Georgia-Pacific Corporation

> "Clip analysis reports give builders a very solid look at the results of their public relations campaigns. We can calculate for you the advertising value your company would have had to pay to gain equal coverage in a magazine or newspaper. When you discover that a press release was picked up in your major daily newspaper and run as a half-page story, that's a significant achievement that should be analyzed and appreciated."
>
> Carol Holden
> Director of Operations
> Burrelle's News Analysis

In addition to actually clipping and sending you the newspaper or magazine story mentioning your company, clipping services can also monitor website mentions, radio and television broadcasts, and other mass media. And, for an additional fee, a clipping service can provide you with a quarterly analysis. This analysis can tell you the location of the placement of the story in the publication, the circulation of the publication, advertising rate value, and specific column inches of the story. Many companies use these details as a quantifiable way to measure the impact of their media relations efforts.

> Interested in seeing what kind of press coverage your competitors are getting? You can arrange to monitor and track any company or competitor through a professional clipping service.

Chapter 2

Getting Started

With some business tasks, getting started is just a matter of taking a giant leap and digging in. Not so with public relations.

A better description would be gathering your breath and putting your best foot forward. Why? Because public relations efforts are extremely visible to a wide variety of audiences–your employees, customers, the media, competitors, and vendors to name just a few. So once you determine it's time to embark on a public relations program, it's important that you do so whole-heartedly, with your best foot firmly placed forward, ready to make an impressive statement in the marketplace.

Assigning the Public Relations Function–DIY PR Versus Hiring a Professional–A Look at Your Options

Just as you have to determine who will lead your company's marketing efforts, you also have to determine who will tackle your public relations function. Determining your public relations goals and budget for your business is a first step. These two decisions will help you determine if you want to add someone to your in-house staff, use a consultant, or hire an outside agency.

The best piece of advice you'll get in this book is to use a professional public relations practitioner. To gain the most out of your investment in public relations, using a professional public relations practitioner is imperative.

Don't mistakenly think that because an administrative assistant has a good personality that she would be good at public relations. Just as you wouldn't put your cousin in charge of sales for your company because he's a nice guy, you also should not put someone in charge of public relations who isn't properly trained to handle the complex tasks involved in properly communicating with a myriad of target audiences.

What should you look for in an internal or external public relations person? Find someone who has been formally trained in public relations, someone who is a dedicated public relations professional. This person should be a strong writer, have excellent organizational skills, and be able to represent your company professionally at all times.

Ideally, you're searching for a person who is aggressive but not pushy. A person who is passionate about public relations and your business. A person who excels under pressure and thrives on juggling an abundance of projects simultaneously. When you find this person, whether you decide to put the

> "What's the secret recipe for a successful public relations program? Employing a professional public relations practitioner with outstanding writing and organizational skills. Look for the person who can just as easily write a speech for you as copy for your website or a sales brochure. Don't skimp. Find a person with formalized public relations training and you'll have a valuable addition on your team."
>
> Shawn Draper
> Vice President of Marketing
> *Woodcraft Supply Corp.*

person on your payroll or work with them as an outside consultant, you'll know that your company is being represented in the best possible manner.

On-Staff Public Relations Experts

Some builders—especially those associated with larger companies—choose to have on-staff public relations experts (or one expert) handle the public relations function. An on-staff professional can be especially helpful because then you have someone on your team who is directly involved in all company operations on a day-to-day basis.

An on-staff public relations expert can support a variety of company needs. Besides working on publicity and promotions, this person can assist with employee relations, sales events, parade of homes participation, award entry submissions, spokesperson training, shareholder relations, speechwriting, special events, and community relations. If your budget allows you to have an on-staff public relations person, your company leaders can sleep a little better at night knowing that they have direct and constant access to a public relations professional who can support them with everything from handling a company crisis to hosting press conferences.

What are the downsides of having someone on-staff? If you're a medium- or smaller-sized company, you may not have the need for a full-time public relations professional. Or, if budgets are tight, your company may determine to go "lean and mean" and not put a public relations person on staff.

> "Having a professional public relations person directly on our staff has always been a comfort and a strong resource for us. As a company, we push ourselves to focus on public relations because we see how valuable it is for our bottom line sales."
>
> Patti A. Grimes
> Vice President
> *Carl M. Freeman Communities*

On-Staff Communications/Marketing People

Although it's always best to hire a professionally trained public relations person to work with the media and coordinate your public relations campaigns, you could have an on-staff communications or marketing person assist with public relations efforts.

You may have this person serve in a support role to a public relations expert or simply determine that you need your limited staff to perform multiple functions.

If you choose to have marketing or communications people handle the public relations function, enhance their efforts with extra training. A person who learns how to write advertising copy doesn't necessarily know how to write a press release. And a marketing person who runs open house promotions may not know how to interact with the media, community leaders, or your company's leadership during stressful times.

It all comes down to expectation levels and training. Don't expect a general communications or marketing person to know how to perform as a public relations professional unless you provide training. Whether it's working side-by-side with a public relations practitioner to learn the ropes or gaining more formalized training in the classroom, your on-staff marketing and communications people need training and support to professionally perform a public relations role that they may be unfamiliar or uncomfortable with.

Public Relations Interns

Regardless of the size of your business, public relations interns–students looking for real world experiences for credit at their colleges or universities–can be an interesting and valuable addition to your company.

If you have an on-staff public relations professional, that person can supervise the intern and gain additional "arms and legs" for assistance with projects. While interns learn from their hands-on experiences, companies often benefit from the enthusiasm, hard work, and dedication these students bring to the job.

For a builder without an on-staff public relations team, a public relations intern can offer short-term assistance and support. It's key to remember that a student is in a learning position while serving as an intern and so needs supervision. A company with a structured internship program, complete with goals and job responsibilities outlined for the student, will have far more success than will a company that views interns as "cheap labor" for getting a project done.

Interns can perform many tasks for a builder. The following list is just a starting point for ideas on how interns can help your company:

- Research and write press releases
- Assemble press kits for media use
- Research and write articles for the company newsletter
- Proof-read websites and make recommendations for enhancements
- Research and update media contact lists
- Organize photography inventory and images
- Assist with details and handling of special events
- Supervise the company's press clipping service and/or find clips on a regular basis
- Circulate press clips to key management team members

14 *Public Relations for Building Pros*

Figure 2-1. Public relations interns can help builders handle special events, track press clippings and put together press kits.

- Research information for management speeches
- Assist with local community relations programs
- Stuff envelopes for media mailings
- Assist with propping and arrangements at photo shoots
- Create selling pieces (e.g., postcards, stuffers, posters, etc.)

To find a public relations intern, contact local colleges and universities with public relations programs. The department head usually assists with student placements and generally has a structured internship program available.

Another source to consider is the Public Relations Student Society of America (PRSSA). Headquartered in New York City, this organization has several hundred college chapters nationwide with both national and local internship programs. More than 10,000 students are enrolled in PRSSA. These students are majoring in public relations at their schools. They seek internships to provide a basis for experiencing what they're being taught in the classroom.

> Once only available in the summertime, college public relations interns are now available year-round. Many students take a semester off to work full-time as an intern in a learning work environment.

Outsourcing to Public Relations Agencies

When looking outside your company for public relations assistance, it's natural to look to an agency. After all, many builders have worked with advertising agencies to create newspaper, radio, and/or television ads. This relationship with an ad agency has perhaps laid the groundwork of understanding for you about how a public relations agency/builder relationship would work.

There are an abundance of public relations agencies around the world, and many marketing and advertising agencies also offer public relations services. First let's consider public relations agencies. If you're unsure where to find a public relations firm, it can be as simple as looking in the phone book or checking with your local chamber of commerce.

Several professional organizations can also assist you in the search for a public relations agency. You can contact the Public Relations Society of America (PRSA–the parent organization to PRSSA) in New York City for a listing of firms or the International Association of Business Communicators (IABC) in San Francisco.

Another great source for recommendations are magazine and newspaper editors. If you have contacts at any building trade magazine, consumer shelter magazine, or even the home editor of a local newspaper, call them and ask for the names of any public relations agencies they're in contact with whose work they respect. Any credible public relations firm should have strong, respectful relationships with editors.

Ideally, when searching for a public relations agency, look for one that has experience in the building industry, or at least in the functional areas your company is interested in. Treat a public relations agency search and review as seriously as you would the search for any other outside support firm. Here are some tips to make your search easier:

- Determine before the search what functions (such as publicity, special event coordination, community relations, crisis planning, etc.) you will need the agency to perform. Then, seek out firms that have expertise in these areas.
- Go into the search having a yearly budget in mind for the entire public relations function–including agency time and project expenses. Usually the public relations budget is a portion of your annual sales or of the marketing budget. Be ready to openly discuss billing procedures, including the development of a plan that works best for your company. This plan may include per hour charges, project-by-project fees, and/or monthly retainers.
- Ask industry friends for references and ideas on agencies they've come in contact with in the past. Your local home builders association may also have a list of recommended firms for you to consider.
- Seek out firms that specialize in public relations for the building industry. This reduces the "learning curve" an agency might experience when trying to understand the industry and your needs.
- Decide if your company needs public relations support from a local agency or if geographic location is not a factor in your search. Smaller builders who have distinct regional focuses may find a public relations firm in the same geographic area best suits their needs.

- Be ready to share information on your company, your public relations goals, and your overall marketing objectives with agencies you contact. This will help them better understand your needs and determine if they would be a good match for your company.
- Make certain that the agency you're considering does not have other conflicting clients. Although it's good to have a firm specializing in the building industry, it's not advantageous to have your designated public relations firm promoting other builders in your competitive market.
- Request and review references carefully from a public relations agency.
- Discuss the background and personalities of the person (or people) the agency will assign to your account to make certain you have a good fit. Determine early in the conversation if this person (or people) has a personality that will fit with your team and appears to be dedicated to long-term employment at the agency.
- Get it in writing. Once you've selected a public relations agency, make certain to have a formalized, written contract clearly spelling out services to be provided and compensation.

> "Our firm specializes in handling the media relations and marketing communications needs of building product manufacturers. We've worked with leading manufacturers of windows, doors, vinyl siding, kitchen cabinets, skylights, wood trim, range hoods, home ventilation, ceramic tile, ridge vents, storage systems, interior paneling, construction adhesives and paints and stains . . . just to name a few. This gives us tremendous experience with the special needs of this industry. For our clients, one of the best advantages to hiring a firm like ours is that there's a short 'learning curve'—we know the industry and can jump right in to support them."
>
> Cary Griffin
> President
> Griffin & Company, Inc.

Just as with any outside organization, it's important that a public relations agency be "kept in the loop" on company information. Public relations people should not be limited to a "need-to-know basis" because they need to know, understand, and communicate a wide range of information about your company. Assigning a direct contact at your company to the agency and keeping the agency personnel updated with constant information will go a long way in creating a successful relationship.

Independent Public Relations Consultants

Eager to gain outside public relations support for your company but not too keen on the idea of getting involved with a large agency or firm? Many builders today rely on freelance public relations consultants—sole practitioners that you can hire on a long-term basis or for just one project.

Your search criteria for a public relations consultant should follow somewhat the same flow as the tips given for finding an agency. However, as a consultant is more independent, you need to recognize that you're "putting all your eggs in one basket" when relying on a consultant. Although some consultants are networked with other support marketing and public relations people, others are sole practitioners and "do it all" themselves.

Determining your needs and budget will help you determine whether to search for a full-scale agency or the more individualized attention you can get with a consultant. In today's business environment there are a growing number of individual public relations consultants available who are dedicated to working just in the building industry. Some of these individuals were downsized from corporate positions at building product companies while others had an entrepreneurial spirit and decided self-employment was their goal.

> An independent public relations practitioner may be the ideal solution for builders of all sizes. Selecting the right person may mean more opportunities to network within the industry with their other clients, being able to co-op community projects, and gaining the expertise of a full-time person while only paying for the work you need. Having an "on call" public relations person can bring added peace-of-mind to many builders—especially during stressful times and in emergency situations.

One of the benefits of finding a public relations consultant with building industry experience is that the person is generally well-connected—to the trade and consumer media, industry and association professionals, and to others in the industry. This can be hugely beneficial to your company.

Once again, finding these independent public relations consultants takes research through PRSA, industry associates, and association contacts. You'll want to determine a financial arrangement that works best for both your company and the consultant. Also plan to have a contract and regular review process in place to add structure to the arrangement.

How much can an outside public relations consultant or agency actually accomplish on your behalf? The sky is the limit. While some builders choose to use their outside support team for limited activities like community relations or media story placement, others rely on outside public relations professionals for everything from special events to spokesperson training to employee relations programs.

Prioritizing Public Relations Goals for Your Company

Just as with any company function, public relations must have measurable goals. How do you determine those goals for your company? Start with your business mission statement and your key marketing objectives. Public relations efforts should complement other efforts underway at your company, not drive off in a totally different direction.

For example, consider a builder's

> If you have just a limited budget for public relations activities, it may be best to start by hiring an independent public relations consultant to work on a project-by-project basis.

> Always create your company's yearly marketing goals and objectives before working on your public relations plan. Your company's public relations programs should support and complement your overall business and marketing goals.

mission statement that focuses 100 percent on attracting the attention of move-up home buyers. The marketing objective is to influence as many move-up buyers as possible to purchasing his homes.

In this case, the public relations goals should be tied in with education–promoting the various benefits of move-up purchases to targeted audiences. Public relations efforts could involve the creation of stories on when it's most important and valuable for home buyers to consider a move-up purchase, the financial benefits of move-up purchases, and/or profiling the ideal move-up home buyer. This complementary strategy is important so that all company elements are working in unison.

Although a central theme such as education can be a company rallying point, it doesn't have to be the sole goal. Public relations allows you to stretch your wings a little.

The builder focusing on move-up target audiences may want to put a more human side on the company's image. This could be a goal of the public relations function that gets achieved by profiling successful move-up homeowners in a series of case study press releases. Or the public relations team could create a qualifying questionnaire for move-up purchase positioning that the sales team could use. Or a television segment could be proposed to the local news station offering tips for move-up home buyers–including the top five preferred move-up features that second-time homeowners look for. All of these ideas support the company's goals.

Where do you start when creating and prioritizing public relations goals for your company? With the bottom line. Identify first how public relations can support the sales and marketing efforts underway at your company and make that a top priority. It may be that your company's biggest marketing effort this year will be participating in a parade of homes competition sponsored by the local home builders association. If that's the case, then top priority strategies should be developed to provide public relations support to gaining awareness of the show home before, during, and after its opening. Start with the groundbreaking and don't stop your public relations efforts until the home is closed up and sold!

> As a builder, you're a "public figure." Remember that you're always "on" and whatever you do in public and in private reflects on your image. That means you can't go to a baseball game, get drunk, and swear at the umpire without considering that a potential customer may be sitting behind you in the stands. Your actions—on the job and off—impact your reputation in the community and can influence sales of your homes.

Identifying Your Target Audiences

If you're like most builders, you can easily identify your target audiences. You know

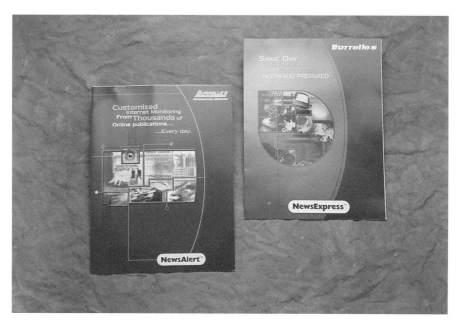

Figure 2-2. Special companies, such as Burrelle's Information Services, exist to help their customers identify and reach key target audiences.

the type of people you're building homes for–their income levels, desires, and needs. You may have the focus down extremely tight, or you may have a wide open net to "catch" almost anyone interested in a home.

Once again, the public relations function gives you the opportunity to support your marketing efforts. If you're a home builder focused on luxury homes, you may identify your top target audience as consumers with a combined dual income of $350,000 or more. Great, you know who you're looking for. Now let your public relations professional come in and help you reach those people with your message.

There are special media outlets that target high-end income couples. Aside from the consumer magazines dedicated to homes, consider the higher end lifestyles of luxury homeowners. Your public relations person may be able to place stories in golfing, yachting, landscaping, or even travel magazines. It's the job of the public relations person to find the different niches, create newsworthy stories, and get the publicity that will result in sales for your company.

Plan the Work and Work the Plan

Whoever said, "Ya gotta have a plan" was so right. Most builders

> When identifying target audiences for a public relations person, share as much buyer preference information as you possibly can. The more audiences you can identify, the harder the public relations person can work to gain multiple story placements and create "matches" of your company information with your target audiences.

already have a business and/or marketing plan, but it's equally important to have a separate public relations plan. Ideally the plans should complement each other yet have differing strategies and tactics.

Planning ahead is critical in public relations. Whether for an open house, a groundbreaking for a new development, or a major company announcement, a plan is the map that gets the job done correctly. Similarly, it's important to actually "work the plan" rather than letting it sit and collect dust for a year. There may be alterations and additions to the plan during the year, but the public relations plan should be the basis for the company's direction and goals.

As many students are taught in college public relations classes, there's a popular formula that can help any builder implement the elements of a public relations plan. The formula is RACE and it stands for:

Research
Action
Communicate
Evaluate

Simply put, public relations students are taught to research a project or idea before implementing it. This would include determining the problem or situation to be handled. Next, take action. Develop a program on how to manage the situation. The third step, communication, deals with executing the strategy and conveying it to the identified target audiences. Finally, in the evaluation step, it's time to review the results of the strategy and see if the goals were reached.

How can you use the RACE formula to help your company? Easily. Consider the public relations person who is asked to help a builder introduce a subdivision. This is the builder's third development in the same community, and the builder is looking for a "fresh, new look" to serve as a "hook" to bring in potential first-time buyers.

Following the RACE formula, research must first be done on potential competitive subdivisions in the area, on how the new development measures up against the competition, on what sets the builder's subdivision apart, and on how the company can best position itself in the marketplace. Next, an action plan is developed. More than likely this action plan will include press releases to primary and secondary media targets, structuring of special events (such as a ground breaking), development of virtual reality tours on the builder's website, working with manufacturers to gain their support for products selected for the model homes, and many other action steps.

When it comes time to implement the plan, the public relations person has stepped up to the communications level. Materials are sent to the media, interviews with the builder are arranged for radio shows and newspapers, materials are developed and sent to targeted audiences (such as apartment dwellers), realtors are informed on the new development, and perhaps there's even a contest to name the development! During the next several months, while still implementing some strategies the public relations person can get a good evaluation perspective by monitoring the press coverage received, the willingness of the media to interview the builder, the number of visits tracked on the website (to view the virtual tour), the number of calls received at the sales center, and how many realtors visit the new model home. The process continues as additional releases are sent out to the media with new photos and story angles.

> "Having a written document to serve as our public relations plan clearly outlines the expectations for everyone involved. There's no guesswork involved. We're all traveling on the same page and working together as a team."
>
> Katie Avsec
> Marketing Manager
> *Carl M. Freeman Communities*

Public Relations Tools

Just because a public relations person works out of an office and not on the jobsite doesn't mean he has no tools to use. Public relations professionals use an array of tools to reach the media, convey messages to target audiences, and impact overall public opinion.

Press Releases and Press Kits

In the media relations arena, the most basic and time-honored tools are press releases and press kits.

> Think of the public relations plan as a checklist of activities that need to be done to get your company to the finish line each year.

Releases are generally created and sent to members of the media when a builder has news to convey, such as the opening of a new development, the announcement of a partnership with an area architect, or winning an award.

Press releases follow a standard format. Editors expect to see the "five W's" in the first paragraph to give them a summary of the most significant information you want to share. The five W's include who, what, where, when, and why. As an example, consider this opening paragraph in a press release from a fictitious company, Brandywine Builders.

Brandywine Builders Unveils Plan for New Subdivision

HARTWELL, GA–Brandywine Builders, an award-winning builder of homes in the north Georgia area, has announced the groundbreaking of its newest subdivision, Friendship Court, scheduled for June 1, 2003. Located on the banks of Lake Hartwell and covering more than 25 acres of lakefront land, 50 homes at Friendship Court will be constructed using universal design products and styles for the easiest lifestyle possible for retirees.

This opening paragraph clearly covers the 5 W's:

Who—Brandywine Builders, an award-winning builder
What—has announced the groundbreaking of a new subdivision, Friendship Court
When—June 1, 2003
Where—on the banks of Lake Hartwell
Why—to create easy lifestyle homes for retirees

The opening statement also includes specific details on the number of homes planned and a description of the "universal design" so that editors immediately grasp the scope of the project. Although the rest of the press release goes into more detail on the planned development, the opening paragraph has given the editor a summary while capturing his attention.

There are other key elements to a press release. It's important to include the name of a contact person (usually the public relations or marketing person) and communication information (such as phone number and an e-mail address) so that an editor can reach your company if they have questions about the material conveyed in the release.

Other standards of press release content include:

- An informative and catchy headline

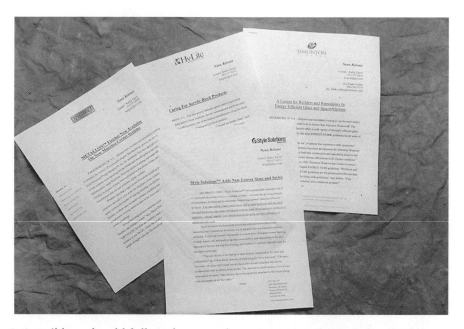

Figure 2-3. Builders should follow the same format as manufacturers and other companies when issuing press releases—all the critical 5 W's go in the first paragraph.

- Double spacing the release
- Never breaking paragraphs from one page to the next
- Quotes from company leaders or third-party credible sources with full title information on the quoted person. Usually releases include quotes from not more than two different sources.
- Concluding a release with a standard boilerplate of information regarding your company, which normally includes a short history on the company, explanation of your services, a website address, and a phone number for reader inquiries
- Either the inclusion of photography with the release or information at the end of the release as to what photography is available to the media and how they can obtain it

Look again at the Brandywine Builders release, this time in its entirety, to identify the key elements described above:

News Release
Contact: Kathy Ziprik
123-456-7890
kziprik@aol.com

Brandywine Builders Unveils Plan for New Subdivision

HARTWELL, GA–Brandywine Builders, an award-winning builder of homes in the north Georgia area, has announced the groundbreaking of its newest subdivision, Friendship Court, scheduled for June 1, 2003. Located on the banks of Lake Hartwell and covering more than 25 acres of lakefront land, 50 homes at Friendship Court will be constructed using universal design products and styles for the easiest lifestyle possible for retirees.

Friendship Court will feature seven different home plans by SLS Architects, located in Troy, GA, that are specifically designed with the needs of young retirees in mind. Universal design features such as barrier-free showers, extra-wide interior doorways, and low thresholds will be included in the homes. Each home will have easy access to the lakefront with moving flat sidewalks to transport people and items from the home to dockside locations.

"We're extremely excited to offer residents of Hart County the unique design elements in Friendship Court," according to Butch Dring, president of Brandywine Builders. "Retirement living will be easier and more exciting than ever before with the amenities planned for this community. We're constructing an on-site restaurant, pool area, nine-hole golf course, and community greenhouse."

Homes in Friendship Court will start at $350,000 for two-bedroom ranch-style houses. Each home will be located on approximately a half-acre of land and be constructed of completely low-maintenance building products.

"Retirees will not have to concern themselves with interior or exterior maintenance of their homes," says Dring. "Friendship Court will be all about easy living without maintenance hassles. The top-quality construction and landscaping plan will assure homeowners of years of worry-free living."

> Brandywine Builders, based in Hartwell, GA, constructs more than 400 homes for retirees in the Hart County area each year. In 2002, Brandywine Builders was recognized for excellence in construction practices by the Georgia Builders Association and received the "Outstanding Builder of the Year Award" from Hart County Builders Association. For additional information and literature on Friendship Court or Brandywine Builders, call 1-800-123-1234 or visit the company's Internet website www.brandywine.com.
>   Photography: Images enclosed of three new Friendship Court designs, by SLS Associates. For additional images, contact K. Ziprik at 123-456-7890.

Press kits go a step beyond a basic press release. Kits usually consist of a standard or company logo folder so that a variety of materials (such as sales brochures, company fact sheet, photos, and an assortment of press releases) can be included inside. There is no right or wrong way to assemble a press kit. You can provide background press kits that supply an overview of your company for when a media person requests it or specific press kits devoted to different developments your company has constructed.

A basic press kit for a builder should include the following:

- A fact sheet on the company, providing historical information on the company, a listing of company officers, awards and honors received, and contact information.
- A basic company press release. This is an expansion of the fact sheet and can discuss your company's philosophies, its target audience, and details on innovations that make your company stand out from the competition.
- A reference sheet. This would include names and contact information for homeowners (approximately five) who have purchased your homes. These people should be contacted in advance and asked to serve as media information links about your homes. This list adds third-party credibility to your company.
- Photography, either a CD of images, a sheet of slides, a printed sheet of images that can be ordered, special media website information, or other ways that media members can obtain key images of your homes–including both interior and exterior shots.
- Media fax-back sheet, a check-off listing for media members to easily request additional materials from your company such as literature, videos, sales materials, and one-on-one interviews with company leaders.
- Company overview brochure. Resist the urge to stuff every brochure your company produces into your media kit. Include one or two key literature pieces and then list other available pieces on the media fax-back sheet.

> When writing press releases, make sure to follow writing guidelines and rules set forth in *The Associated Press Stylebook and Libel Manual*. Editors universally prefer the formats outlined in this stylebook.

Your general press kits should be updated prior to sending them out by adding your most current home development information or the most recent news your company has announced (e.g., merger with another company, introduction of a new company president, etc.). Include valuable up-to-date information that could assist the editor in creating a story on your company.

> "Just like our counterparts in the print media, in radio we find that the most useful press releases are those that clearly and concisely present the facts."
>
> Tom Kraeutler
> Nationally Syndicated Radio Host
> *The Money Pit* Home Improvement Radio Show

Visual Support: Slides, Photos, CDs, Videos, Transparencies, and Websites

You've heard it over and over again: A picture is worth a thousand words. In the housing industry there is no truer statement.

Have a great image of your new signature home? Then you'll likely get coverage in the local media. Sending a fuzzy image out to the press? Consider it an immediate part of their trash pile.

People like pictures, so magazines and newspapers need good quality pictures to help tell their stories. This means that builders must invest in professional photography on a regular basis to keep their products in front of their target audiences. The good news is that photography you take for public relations purposes can easily have other uses, such as in ads, catalogs, brochures, websites, and flyers. Investment in good photography can pay off nicely for your company.

> When you visit local or national home shows, stop by the press room. Pick up one or two building product manufacturer press kits for products you use on a regular basis. Then, contact the public relations contact for the company and suggest some of your projects for their promotional materials.

There's so much to say about visual support as a public relations tool that it's hard to know where to start. You may feel the same way when you review the options available to you–prints, slides, CDs, websites, large transparencies. What type of artwork should you obtain? What do media representatives really want?

Surprisingly, high-quality duplicate slide images are still preferred by most media contacts. We say

> Always have 10 current company press kits available for the media. Many media members will request overnight delivery of materials to meet a deadline. Having kits ready means not scrambling to meet the short timeframes of the media.

Figure 2-4. CDs of information and photos, like this one prepared by St. Thomas Creations, can quickly educate media people on your company and the support materials you have available.

surprisingly because many people imagine that media representatives are all steeped in high technology and are eager to have CDs and private access websites. Not so.

Here's the secret: Print media outlets (especially magazines) are not as advanced as we may think they are. Plus, they crave ease of operations. If you meet this need, your product is halfway in the publication!

When an editor at a trade magazine like *Builder* receives a press release from you with photos of your new model home and several attached slides, they can immediately hold the images up to the light. Meaning they can make an instantaneous, two-second decision as to the quality of the image and their interest in the subject matter. If they like it, the images may get passed on to a freelance writer who will need the release and information to create the story. Or they may go directly to the art department for layout. Or one of the images may go into the editor's file for an upcoming story while others are passed along. Regardless, your mission is accomplished. Interest level was immediately and effectively raised by sending a slide.

Now consider the same release arriving with a CD. The editor has to put the CD in her computer and download the images. This can take valuable time away from the editor's day. What if there are several images the editor wants to consider for different projects? She has only one CD. It can be a lengthy process to download and send different images to different people. As one editor has said, "If you really want to send out CDs, at least make sure there are thumbnail images of each shot directly on the CD and on the container cover to make life easier."

Let's take this idea a step further. You send out the press release with no images. Instead, you tell the media representative to either visit a private website for image access or call you for images. You may have just lost the editor. When you consider that most editors get several hundred press releases a week, you quickly realize that their time is valuable and that your competitors may be sending out exactly the materials the editor needs, making it easier to use their materials.

The best idea of all is to offer editors a buffet of choices. If your budget allows, send out slide duplicates and CDs (or make a private media website available to the media). Forget about print images unless they're specifically requested. And always be ready with some extra 4 × 5 transparencies for media people needing larger formats of your images.

> Websites say a lot about a company, and these days media members are using them as resources for stories. Take a good look at your site to make sure it has the most updated information possible. Then visit www.freemancompanies.com for inspiration.
>
> Carl M. Freeman Communities is an established resort community developer on the eastern shore of Delaware. The company has constructed resort communities for more than 30 years. This gives the company's public relations team lots to talk about to the media, and they've done so in a wonderful "Press Information" section of the website.
>
> Look at the Carl M. Freeman Communities press information area and you'll find an editor's dream—bulleted information on more than a dozen different aspects of the company, including its history, leadership, community involvement, service philosophy, environmental practices, and properties. Even the smallest building company can take note of the detailed information on this website and duplicate the excellent format for its own website.

By-Lined Stories

With smaller local newspapers and with some trade magazines there may be an opportunity to supply articles on specific topics. It takes time and patience, along with good writing abilities, to gain the trust of an editor to run your company's stories. It also takes the ability to offer unbiased, balanced information.

Publications such as *Shelter, Rural Builder,* and *Architectural West* all work with companies and public relations people to supply expert information to their readers. Generally, a manufacturer receives a topic from the editor of the publication, along with a certain word limit and parameters for the story. And the

> When sending out slide images, make certain to imprint the mounts with your company name and phone number in case the slides get separated from your release. Protect the slides in sturdy, heavy duty slide sleeves so they don't get damaged in the mail.

> "Companies that wouldn't dream of having their homes represented by an unintelligible sales rep wearing a ripped T-shirt and sandals think nothing of sending fuzzy, off-color print photos with their press releases. On the pages of a magazine, photos are the best tool for convincing readers that a home is attractive and well-constructed, while the text should explain the features and benefits of the home. Photos speak to the quality and creativity of your home, and, for readers, a poor photo equals a poor product."
>
> Robert Wilson
> Editor
> *Better Homes and Gardens Home Planning Ideas*

editor always has the right to change the content of a submitted story to meet the publication's needs. This can also happen to builders who supply stories to newspapers and magazines.

Why go through the time and effort to supply by-lined stories? Because they gain your company credibility and awareness. Consider *Architectural West* magazine, which reaches almost 20,000 west coast architects with each issue. In the front of the magazine there is always a "Fenestration" column, generally 1,000 words (with photos) on a timely topic. Having the president of Simonton Windows, John Brunett, as the authority who is presenting information to architects on a variety of topics is a major achievement for Simonton and positions the company positively.

Mr. Brunett's column, which has run in every issue for over a year, never mentions Simonton Windows by name. Instead, he gives information on everything from window trends in residential homes to juggling window shapes and sizes in architectural designs. The underlying message is clear to the readers. Simonton Windows is a leader in the industry—selected by the magazine to share valuable information. This positioning with architects couldn't be better for Simonton Windows and serves as a major part of their architect marketing initiative.

Although it's always nice to see a company president's name in a magazine as the author of a story, don't shy away from opportunities to supply articles without a by-line. Creatively written to look like the editor of the magazine wrote the piece, a placed story can have just as good an impact—especially when it quotes a key company spokesperson.

> It doesn't matter who researches and writes a by-lined story. Unless it's a technical piece written by an engineer or specialist, always list your company president as the author to add stronger credibility to the company as a whole.

Industry Trends Stories

Like clockwork, the calendar pages turn each year from December to January. With equal predictability, consumer and trade media are interested in what will be "hot and new" in the coming year.

Trends stories are crystal ball projections of what industry experts expect to

Figure 2-5. This bylined story in *Architectural West* magazine brings tremendous exposure to Simonton Windows with architects—one of their key target audiences.

occur in the future. The idea is that reliable sources use their research, gut feelings, and best analyses to suggest what will happen to our industry.

Many public relations professionals find that issuing trends press releases and stories at the conclusion of a year—oftentimes as early as October—gain fast and solid results. Again, these type of visionary stories position a builder as an industry leader while allowing you to weave in your specific company messages. Rarely are the stories just about one builder. A valuable trends story or press release will quote industry experts and look at the "big picture" to make the piece insightful.

Trends stories give a glimpse of the future to readers and provide your company

> After agreeing to supply a custom-written story to a publication, go the extra step. Supply appropriate images to support the story with captions that subtly reference your company and/or products.

> "Trade magazines such as ours rely on balanced information from industry professionals. We often run articles, provided they are unbiased and well written, submitted by public relations people for their companies."
>
> Ryan Reed
> Editor
> *Rural Builder*

30 *Public Relations for Building Pros*

Figure 2-6. Industry trends stories—whether on product advancements or changes in consumer buying habits—are always popular with industry publications.

the opportunity to position itself for the new year. Some variations of topics might be:

- Homeowners to Seek More Low Maintenance Products in Coming Year
- Builders Predicted to Use Larger Numbers of Engineered Lumber Products in Near Future
- Hot Metal Trends for the Future
- First-Time Homeowners Want More Closet Space
- High-End Upgrades Top Consumer Wish Lists
- Homeowners Warm Up to Energy-Efficient Windows
- New Advancements in Product Research Spur Growth Anticipation for Mildew-Resistant Products
- Educated Consumers Demand More Information on Home Construction
- Basements' Hottest Trend for Second-Time Buyers
- Environmentally Friendly Products Top Consumer Wish List for New Year

Media Binders

For large builders with a wide selection of properties or an abundance of media materials, the media binder is a good way to offer information to editors. Some builders create this "one-stop shopping" type of press kit in a three-ring binder. The binder contains a variety of materials, professionally written, printed, and created, that can serve as a deskside reference for an editor.

Top-quality media binders have a shelf life of no more than two years. After this time, company and product information should be updated and new photos offered.

> If you're going to produce a media binder, invest the time and money necessary to make it professional looking.

Media binders can include color pages of photography with fax-back forms for the media to request images on CDs or in slide format. Some companies also offer electronic versions of media binders on special websites designated especially for the media or on CDs.

Media binders can be both costly and time consuming to create. However, when done with the media's long-term needs in mind, they serve as an excellent resource for editors' day-to-day needs.

Deskside Briefings

The theme song for some public relations people undertaking national deskside briefings should be "On the Road Again." For public relations people dealing with builders in more localized markets, the travel demands aren't as great, but the preparation remains hefty.

As simple as they sound, media deskside briefings involve a great deal of time to schedule meetings with key media contacts and give presentations at their offices. Getting a commitment from editors to meet with you can be achieved if you offer one of the following at the proposed meeting:

Figure 2-7. Media binders are an excellent way to present all your company information and public relations materials in one comprehensive unit to media people.

> "We get mountains of media materials every day. The materials that have lasting value are those that address what is important to our builder readers rather than just to the manufacturer. Explain the builder benefit up front and you'll capture my attention."
>
> Heather McCune
> Editor-in-Chief
> *Professional Builder*

- New information, literature, and/or photography
- New trends research related to the industry
- A high-level company representative to share new information on the direction of the company and/or trends in the industry

Schedule deskside briefings only when you can bring new information to editors that they will find valuable. Much like a sales call, you should have an agenda before going into a deskside briefing, along with leave-behind items to share. Practice is essential to make certain you can cover all your topics in 30 minutes or less.

Here are the steps a builder and his public relations person should complete to host a successful series of deskside briefings:

- Identify the agenda items you want to discuss with the media. List materials you will take on the tours, including leave-behind materials and samples.
- Customize the above list for both trade and consumer media, making sure to tailor your presentation for different media types.
- Line up company participants for the deskside briefings.
- Identify several cities hosting a concentration of your targeted media.
- For each city, map out the targeted media contacts, determining how many meetings can comfortably be scheduled in a day including travel time. Maximize the day by suggesting breakfast, lunch, and dinner meetings.
- Contact editors individually about three weeks in advance to set up meetings; then e-mail or fax confirmations of time, location, and date of meetings, along with a "sneak preview" of agenda items.
- Reconfirm meetings with editors two days beforehand.
- Prepare and ship any materials needed in advance to your hotel if traveling out-of-town.
- Continually check your voice mail while on the road in case editors make last-minute cancellations.
- At your deskside meetings, set the pace to make sure you cover all agenda items during the allotted timeframe.
- Take notes during your meetings and follow-up with personalized letters and any materials requested by the editors.

Subdivision and/or Show Home Tours

Some builders believe that the best way to understand their "product" is to actually see it. This premise can translate into creating videos, brochures, and CDs that are sent to media representatives and/or getting them to visit your projects.

As with other elements of public relations, consider that the media may not be your only target audience for tours. It may be beneficial to explain and showcase your homes–and the construction steps–to customers, salespeople, and realtors. This decision may lead you to produce a short video or CD and make it available to multiple audiences, including the media.

> In walkable cities such as New York or Chicago, wear comfortable shoes and take your literature and press materials with you in suitcases on wheels.

The ultimate step in educating someone on your homes may be to actually have them visit your homes. This can be arranged as one-on-one tours or in small groups. If you proceed in the small group direction, it may be beneficial to mix up the dynamics of the group by including several editors in with a group of dealers, building product manufacturer reps, past customers, and/or land developers. The value of the session increases greatly for the media in this case because they can make contacts with not only representatives of your company but also industry people and consumers.

When inviting media representatives to your homes, recognize that this is an investment of time and expense not just for your company but also for the media. Because many publications are short-staffed, editors find it difficult to get away from the office.

Consider these tips for tours of your homes when the media are involved:

- Arrange to have your most knowledgeable and well-spoken people lead tours.
- Allow plenty of time for questions; you may even want to assemble a team of internal experts to guide people through your homes.
- Give media people plenty of time to "explore" the project on their own so they can come to you with questions after getting a feel for the home.
- Have refreshments available.
- Assure that the homes are thoroughly cleaned before the visit.
- Designate a working restroom for your guests.
- Take the opportunity to explain about the products in the home and why they were selected.
- Have the interior designer on hand to answer questions.

> When considering hosting a media tour, ask yourself what is so different about your projects that editors should consider investing their time in coming to see. What is unique about your homes? What "walk-away value" is there for an editor who tours your homes?

Media Placement Services

If you're a large builder, you may need help spreading your message throughout a specific geographic region or nationwide. Don't worry. More than a dozen large media placement services are available throughout the country that specialize in everything from broadcast media announcements to extremely targeted messages to specialized audiences.

One of the easiest ways to identify these sources is to request a copy of *"The Green Book"* from the Public Relations Society of America (212-460-1426 or www.prsa.org). This comprehensive booklet is a guide to public relations service companies covering a wide variety of topics from media training to webcasting to annual report design to clipping services. Under the categories of "press release distribution" and "newswire services" you'll discover companies with national and regional offices available to assist you. Here's a glimpse at what you'll find:

- Business Wire–800-221-2462–Specializing in business distribution of news to traditional and online media, including more than 16,000 online database sites worldwide.
- Media Distribution Services–800-MDS-3282–Promotes availability of more than 250,000 editorial contacts at more than 50,000 print and broadcast media in North America, plus all daily newspapers worldwide.
- Metro Editorial Services–212-947-5100–Specializes in themed sections (including spring and autumn home improvement sections) that are sent to more than 7,000 daily and weekly newspaper editors nationwide.
- North American Precis Syndicate (NAPS)–212-867-9000–Sends releases, scripts, and videotapes to all media outlets nationwide. Includes themed sections going to 10,000 newspapers with each mailing.
- PR Newswire–888-776-0942–Specializes in fully integrated communications for broadcast, print, and the Internet–everything from consulting to production to distribution and reporting.

Although these services are relatively expensive, oftentimes their reach and ease of operations can make the investment worthwhile. Other times, especially during a crisis mode, the ability of these service companies to work quickly and effectively as a back-up team can be extremely valuable.

Company Literature

Oftentimes builders will introduce pieces of literature–whether small tri-fold brochures or oversized coffee table full-color books–that are designated for various target audiences and promote their homes. The types of literature that fall into this category include:

- Sell sheets
- Sales catalogs
- Flyers
- Design and idea books
- Subdivision directories

As these pieces are introduced at your company, determine if they can help educate the media on your homes. If so, send them with press releases.

When introducing a new piece of literature to potential customers, send the literature with a press release to local media. Include details such as how the literature can be obtained, the cost of the piece, and the key prominent features of the material. For example, Lighthouse Builders may create a pamphlet entitled "Things to Consider When Contemplating a New Home Purchase." The 12-page brochure would be free to consumers. By sending it to the media, they could promote the availability of the piece and generate interest from potential home buyers.

Figure 2-8. Remember to maximize your company literature by sharing brochures, sell sheets and flyers with your key target media members.

Budgeting for Public Relations

Oftentimes smaller builders worry that they don't have the dollars to host a public relations campaign. What they don't seem to realize is that they can't afford *not* to invest in public relations.

Public relations efforts don't have to break the bank. But they do have to take place in order to elevate your business, stay at the same level with (or better yet, lead) your competitors, and prepare your company for unforeseen situations. This said, the ability for a company to have an aggressive and successful public relations program should not be tied to dollars and cents.

A local builder with as little as $3,000 to invest in public relations can make an important splash introducing a new model home to the marketplace with several public relations strategies. A mailing to several dozen local media representatives (including a press release, sales literature, and duplicate slide images) could cost as little as $500 (including the time involved for a freelance consultant

> Some magazines will run an image of the cover of a new piece of literature offered by a manufacturer. Before issuing your press release, make sure to have a quality studio shot in hand of the literature piece.

> Do not play the "stop and go" game with public relations. If you start a public relations campaign, stick with it year after year to achieve maximum benefits for your company.

> "Think of your public relations budget as an investment in the future of your company."
>
> Kim Dres
> President
> Drew Public Relations

to write the piece and your office staff to stuff envelopes and send out the piece). That $500 investment could result in several solid stories running in newspapers and on television stations reaching thousands of potential customers.

There's no right or wrong way for a builder to determine a public relations budget. Some builders dedicate 10 to 15 percent of the marketing budget to public relations activities. Other companies determine a set dollar figure based on a proposal from an agency or consultant.

Small- to medium-sized builders can expect to invest anywhere from $4,000 to $50,000 per year for an effective public relations campaign, depending on how aggressive they wish to be. Larger builders, with more projects to promote in more geographical areas and with loftier goals, may need to double that number. More aggressive builders have been known to triple or quadruple that amount. The key is that a public relations budget is a necessary part of the marketing function at a company. The more you invest, the greater the potential returns.

Although there is no set ideal budget for a public relations campaign, let's look at a solid $50,000 yearly budget for a medium to large builder with aggressive marketing and public relations goals. With such a budget, a builder can expect the breakdown to look something like this:

Yearly Public Relations Budget

Fee for one on-staff person or yearly support from a public relations agency or independent consultant	30,000
Photography shoots throughout the year	5,000
Duplication of photography into slides or CDs	3,000
Press release mailings (postage, copies, etc.)	3,000
Special event funding	6,000
Clipping service for press clips	1,500
Analysis service for press clips	1,500
Total	**50,000**

Tracking for Success

Public relations results sometimes are difficult to measure. What specifically makes a consumer walk into a model home and get interested? Is it the construction quality? The floor plan? The neighborhood? The landscaping? The reputation of the schools in the area? Reading a story in the newspaper profiling your community? Serving on a community board with one of your employees who bragged about your homes?

More than likely it will be a combination of all these things. The more points in your favor, the more likely you are to make a sale. And, while realtors and your sales team may be hard at work pitching your properties, your marketing efforts can also be hard at work behind the scenes reinforcing their efforts.

Because public relations is a complementary part of the marketing program, there are many times when the results of a campaign cannot be accurately or specifically measured. If your public relations person is challenged with promoting the involvement of your employees in a company blood drive for the Red Cross, you can determine success based on participation levels and on the specific amount of blood donated. But what happens if that same public relations person is asked to research, write, and submit stories on home buyer preferences for local newspapers? Your gut feeling tells you that the information in the stories will impact a potential home buyer, but there's no real way to measure the results on a daily basis.

Measurement Tools and Resources

Fortunately, the results of actual press coverage can be measured in a variety of ways. When you read a story in a newspaper or magazine, see your homes on television news broadcasts, or hear a radio interview promoting your company, there are ways to measure the impact of these public relations messages.

A number of companies, most notably Burrelle's Information Services, Bacon's Information, Inc., and Luce Press Clippings, all offer clipping services. For a monthly and per story fee, these companies will constantly read through and search thousands of newspapers, magazines, websites, radio shows, television shows, and other media outlets in search of instances in which your company and its products are mentioned. Then the clipping service can send you a copy of the piece with as much detailed information as you need–date of the clip, circulation/viewership numbers, position of the piece in the publication or show, and/or equivalent advertising dollars (the amount your company would have had to pay to place an ad of the same size in the exact same location in the publication or on the show).

These clipping reports can be extremely valuable when looking for a way to evaluate the success of a particular public relations campaign or an entire program. The reports can show specific geographical placements and the type of coverage your business is receiving.

38 *Public Relations for Building Pros*

Figure 2-9. Several companies offer press clipping services. Using one of these companies assures that you don't miss any media coverage your company might generate.

Sharing the Good News

What should you do with your clips once your business receives press coverage? Press clips are not just feel-good pieces that should be stacked in a drawer and forgotten. To maximize your investment in a clipping service, make sure to share copies of the clips with your internal and external audiences. Here are a few ideas:

- Send copies of the best clips you receive on a monthly basis to your key realtor contacts. This will give them new information to share with potential customers and it showcases your company's achievements.
- If you have an internal sales team, make certain your sales reps get copies of your clips. They can show them to potential customers and explain how positively your company is positioned in the media.
- Post a selection of press clips on bulletin boards in key areas of clubhouses and meeting areas in your developments.
- Include copies of especially favorable clips in your employee and/or shareholder newsletters.

> For a complete listing of outside companies offering clipping services, consult the Resources Section at the end of this book. When contacting a service, make sure to review all options for clipping and analysis services. Each company offers many "ala carte items" to choose from so that you can tailor your clipping service to meet your company's exact needs.

Getting Started **39**

Figure 2-10. Maximize your press clips by posting them on employee bulletin boards, sharing them with your sales team and creating a "brag book" for your reception area.

- If you get a truly remarkable press clip—such as an interview with the company president in a prestigious publication—have professional copies made. Send them with a letter to your top potential customers.
- Create a "brag book" of your best press clips for people to thumb through in your reception area while they wait to meet company officials.
- Select and copy some exceptionally good clips and send with follow-up letters to potential customers.

If sharing television show results, try some of these ideas:

- Set up a VCR in your sales center and play the tape continuously.
- Notify potential customers in advance of the television broadcast with a postcard mailing.
- Create countertop cards with exciting television show information for display at local dealers and distributors businesses and in your sales center before the airing of shows.
- Send copies of the television show tape to your sales team and key realtor contacts so they can share it with customers.
- Have someone take still photography during the taping of the segment and include the photo with a story in your company newsletter.

> Be sure to check liability and limitations for copying and showing television tapes with the show's producer. Each show has different rules. Some shows, especially those seen on PBS, have strict restrictions on rebroadcast uses.

Chapter 3

Tackling Special Events

Most people define a special event as a unique happening in their lives–a wedding, a 25th high school reunion, a graduation ceremony for a son or daughter. For home builders, special events take on a whole new meaning: the groundbreaking of a major subdivision project, the opening of a new model home, participation in a parade of homes. These are the marketing tools that home builders use–just as surely as they use the tools in their toolboxes–to gain attention from prospective customers.

Defining Special Events for Home Builders

Need to understand more about what makes an event "special" for a home builder? Whenever you can leverage an activity or occasion to generate publicity and customer attention for your projects and/or your company, that's a special event in the making.

Special events are important for builders because they give them something to "rally around" and draw attention to their business. A special event–whether it's solely sponsored by your company or you're simply participating in someone else's activity–allows you to say "look at me and my work!"

To maximize a special event, you can typically send out press releases to the media, pay for advertisements in the local newspaper, send information to past and potential customers, include updates on your website, and use endless other strategies to get the word out.

Imagine your business without participating in (and maximizing!) special events. Fairly boring–for you and your customers. You'd be staring at 365 days a year of uninterrupted building construction. Special events let you and your company jump in front of potential customers to attract attention to the projects you work so hard on.

Logistics of Special Events

An event is not "special" until you make it so. Look at Halloween. Is it special unto itself? Not really. Only once you have children in costumes, decorate the house, and hand out candy does the event become special for everyone involved.

It's much the same for a builder's event. The anniversary of your company completing its 1000th house can slip quietly by unless someone marks the event with a special celebration. Occasions like these should be leveraged to maximize your potential for media and customer attention.

The good news is that every company can turn a relatively ordinary activity into a newsworthy special event. The flip side—or the bad news—is that it takes an investment of time, effort, and coordination to do so. That's where your public relations expertise and team players come in.

For starters, follow these simple steps when contemplating a special event:

1. Identify the idea for the event. Bounce the idea off some folks and determine if everyone believes it has the potential to attract attention to your company.
2. Create goals and objectives for the event.
3. Determine who will coordinate all the details of the event.
4. Set a realistic budget to achieve your goals.
5. Create a written action plan, including a timeline for success.
6. Identify a listing of your needs and assistance areas.

The last item is an extremely important point. Oftentimes builders think they need to do everything on their own, but they don't. A terrific example of manufacturer support is the Owens Corning System Thinking Show Home special events introduced in local markets in 1996. The company provided builders with binders full of ideas and step-by-step programs for hosting special events in their local markets.

Owens Corning supported their builder customers by providing ideas such as a "Puttin' On the Pink" Party, VIP previews of special homes, parade participa-

Figure 3-1. At this Owens Corning System Thinking Show Home, the builder in Ohio was all smiles—a grand opening special event for the home brought more than 3,000 people through its doors in one weekend.

tion support, and home show exhibits. By working together with Owens Corning, many builders found it easy to customize their special events to capture attention in their local markets for a variety of special events.

Although no two special events are really identical, there are similar facets of special events that should be considered universally by all builders. Whether you're planning a barbecue cookout for a vocational tech class that worked on your jobsite for the summer to gain hands-on experience or it's the grand opening of a special show home, remember to consider these important logistical factors:

1. **Determine a date and time that works for everyone.** Plan far enough out for a special event so that you have time to set up all the details. Keep away from major holidays and traffic rush hours. And remember to publicize the times and dates so that everyone knows when to attend!
2. **Create written invitations.** Cover all the key points of information–when the event is, who is hosting the event, where it is (including directions), what type of event it is, why it's being held, who is invited, and attire (casual, business, semi-casual). Also, if there is any special RSVP information or instructions for the event (such as bringing a donation for a charity), include those details on the invitation.
3. **Create the invitation list.** Spend lots of time to make sure that all key target audiences are identified and invited to your event. Make certain you consider potential and past customers, realtors, working media members, VIPs, local government officials, key suppliers, and subcontractors. Depending on the event, you'll also want to determine if the general public and your employees and their families should be invited.
4. **Hire a photographer.** Hire a professional photographer (not your secretary with a point-and-shoot disposable camera!) to shoot the event. Use the pictures afterward in company literature, on your website, to send with post-event press releases, and as mementos for attendees.
5. **Handle the details.** Caterer. Florist. Audio/visual experts. Entertainment/musicians. Decorations. Signage. Parking. All these elements must be coordinated and arranged far in advance to ensure a successful special event.
6. **Expect the unexpected.** Plan in advance for bad weather and other unforeseen problems on the day of your event.
7. **Plan the program.** If your special event involves speeches or a program, map it out. Determine beforehand exactly who will make presentations, in what order, and for how long. Keep it short and you'll have the undying thanks of everyone involved.
8. **Make the day memorable.** Decide if you're going to have gifts for attendees or company literature or displays available. Consider location for all these items, plus nametags, sign-in books, and other special elements to keep track of who attends the event.
9. **Sweat the details.** Special events mean making sure EVERYTHING is considered beforehand. Restroom facilities and extra toilet paper–along with

Tackling Special Events **43**

Figure 3-2. Special events require lots of details . . . right down to the flowers, goodie bags and bagels at this open house on a trade show floor.

Figure 3-3. Invitations to special events can take many forms, including these simple flyers. Remember to send out invitations at least 4–5 weeks in advance for a major event.

A special event with lots of logistics involved is the annual Home Expo presented by the Greater Atlanta Home Builders Association. Hosted by Woodmont Golf and Country Clubs (a John Wieland Signature Neighborhood), the 2001 event involved a parade of six unique homes created by different builders.

During the one-month Home Expo, there were designated Family Fun Days, Realtor Day, Zoo Atlanta Day (one of the beneficiaries of the ticket sales), Ladies Day, a Special Decorating Extravaganza, Artist's Trunk Show, and Sale and Date Night. Each event specialty day was a "special event within an event" that had to be organized and promoted. From the coordination of street hockey rinks and celebrity appearances by the Atlanta Thrashers to an appearance of Zhu Zhu the panda, Home Expo 2001 was crammed with interesting activities and media events.

The homes built by Colonnade Custom Homes, Beardslee Custom Homes Inc., Chathambilt Homes, LLC, Pete Witalis Homes, Inc., Haven Properties, and John Wieland Homes & Neighborhoods received thousands of visitors—and mountains of free press—during the month-long celebration. By linking up with Zoo Atlanta and Cherokee Arts Center as beneficiaries of Home Expo 2001, the planners found willing support from strong organizations that could offer interesting attendee benefits and coordination assistance.

who will monitor these items at the event. Continuous removal of trash. Securing any special permits needed. Safety and security—do you have fire extinguishers and first aid kits on hand in case of emergencies?

10. **Sweat the details some more.** Triple-check everything the day before the event. Call all your key contacts (such as the caterer, photographer, VIP speakers, etc.) to make sure they're still on board. Check with your insurance carrier to make certain you're covered no matter what. Have you considered walkways and facilities for the physically challenged? Did you decide what attire all your employees will wear? Does everyone have business cards to distribute? The details are endless but that's what makes a special event successful!

Timeline for Success

This is the easiest and shortest chapter you'll ever read. Plan early. Think you're fine planning a special event three or four weeks in advance? You're not. For a successful special event, start planning three or four months in advance.

You can never start planning too early for a special event. Is your company now in its 23rd year of business? Start initial plans *now* for a major 25th year anniversary celebration. Planning is the key part of hosting a successful special event. The earlier you start planning, the more successful the event will be. It's that simple. Plan early.

Checklist of Winning Special Event Ideas

What type of special events can you get involved with? Here's a starting checklist. You may want to sponsor an event yourself, investigate working on one with your local home builder association, or even get together with some area businesses to create an event that benefits everyone involved.

Special Events for Home Builders:

- Street of Dreams participation
- Parade of Homes involvement
- Sponsoring an educational session (such as "Tips for Second-Time Home Buyers" or "What to Ask Your Builder" classes) at your offices, a community facility, or real estate office
- Grand opening of a new model or show home
- Creation of a charity show home
- Co-sponsorship of a Ladies Night at a local dealer for single women home-owners
- Hosting a party for everyone involved in a Habitat For Humanity project to celebrate the successful completion of a build
- "We Love Our City" clean-up day in areas surrounding your home projects, involving your past customers, employees, suppliers, and local officials
- Ribbon cutting or groundbreaking on a new development
- Kick-off day at the start of a new community volunteer building project, involving you and your employees
- "Ask the Builder" website chats
- A major anniversary for your company's business
- Hosting a one-year anniversary for all residents for a subdivision or project you constructed
- Subdivision holiday celebration of lights. Work with all your past customers in a subdivision to create a "Street of Lights" for everyone to enjoy that showcases the beauty of their homes.
- Yearly appreciation celebration for all your customers
- The completion of your 100th house, or 1000th

If you're planning a major special event, like the 50th anniversary of your company or a huge parade of homes, here's an ideal timeline for your event:

Nine months in advance:
- Develop parameters of the event
- Set budget
- Set location, date, and times
- Set theme
- Determine any speakers needed

Six months in advance:
- Put together a coordinating team
- Start public relations and marketing efforts
- Secure locations if needed
- Consider logistics for event
- Confirm speaker, sessions, and larger details
- Develop promotional pieces

Five months in advance:
- First mailing to target audience
- Continue work on logistics
- Add event information to website

Three months in advance:
- Send out additional promotional mailings
- Send press releases to media
- Meet with coordination team to review all details

One month in advance:
- Send confirmation letters to VIP attendees
- Extensive review of all logistics
- Full force marketing and promotional efforts

Maximizing Special Events

Whew! The party's over, the special event is finished, and your work is done. Right? Wrong!

Why stop when the last bag of trash is removed? To truly maximize the investment of your company's time and money in an event, let it work for you *after* the event is over. Here are some ideas to really maximize the "afterglow" of your special event:

- Send pictures with a post-event press release to local newspapers to gain publicity.
- Post pictures and a recap of your event on your company website.
- Send your post-event press release and photos to your local home builder association and key trade magazines for use in their publications and on their websites.
- Send thank you notes to all attendees. For VIPs, send a memento, such as a photograph of them at the event, as a way of saying thanks for participating.
- Make certain to thank the people behind the scenes. If there are special people in your company who worked extra hard to make the event a success, give them a gift certificate to a local restaurant and encourage them to relax with family members.
- If your event was truly original and interesting, enter it in a state, regional, or national builder awards and/or marketing competition.
- Include information on the special event in any publications you may produce (such as customer newsletters).
- Use a terrific photo from the event and create a "Wish You Were Here" postcard. Send it to key people on your invitation list who couldn't attend.
- If your event raises funds or collects items for a charity organization, make sure to request a letter from the recipient organization acknowledging your participation. This letter should be framed and posted in your office and shared on your website with customers. This third-person credibility makes your company look great!
- Create a memory wall of photos (or a photo album) from the event for your sales office.

> Don't forget to invite local officials, anyone connected with financing your homes, realtors selling your homes, and your suppliers on the "A" list to every special event you host. They might not always attend, but they should always be invited.

Chapter 4

Spotlight on Media Relations

As a builder, your company offers one product for consumers to purchase: homes. The ability to get free publicity for your company's homes is one of the most valuable aspects of public relations. However, just as with any marketing element, successful story placements don't just happen. Strategies must be in place for achieving solid media placement. Relationships must be cultivated with members of the media. Product messages must be developed.

To get started, recognize that the first element in developing good media relations for your company is to understand the media and identify the right media for your message. A builder selling multi-family low-income housing units will have different media messages than will a builder selling high-end luxury homes. Developing and correctly placing your media messages is critical to your success.

Understanding the Media

Just as you need to plan a sales strategy to reach potential home buyers, you need to do the same for media members. Understanding the different types of media members–and what their needs are–is critical.

Your first step in understanding the media is to determine your targeted media audiences. Are you interested in impacting consumer or trade media? Home improvement radio and television programs? Local or national media? All of the above?

Once you determine your company's target media, you need to read (or watch and listen to) and study the media they produce. To pitch your company information to a publication, you need to understand the style of the magazine, the types of stories they run, and their needs. Same with television and radio. It's useless to offer yourself as a builder expert to a television show that never uses outside spokespeople. By monitoring the media you gain a better understanding of the type of information they require, allowing you to personalize your materials.

The Five Ws of Media Relations

When working with the media, remember that they're always interested in the Five W's when gathering information for a story. As explained in the press release description part of Chapter 2, the Five W's include Who, What, When, Where, and Why.

These Five W's form the basis on which the media determine if a story will

Figure 4-1. Contractor Bruce Rosenthal understands the media. That's why he wore his "Handyman Life Member" jacket to an installation photo shoot for miterless corners for Style Solutions. His quick thinking of wearing the jacket allowed the manufacturer to gain a four-page story in *Handy* magazine on Bruce's installation project.

interest their readers or viewers. Let's take two examples to review how this theory works.

In the first situation, HHS Construction company, a medium-sized builder of starter homes, would like some publicity for their new development. So far they know that the development will be located in Fort Lee, New Jersey, and that groundbreaking will occur on June 1. Do they have enough to create a press release? Yes, but just barely. The lead graph to capture the media's attention might look something like this:

HACKENSACK, NJ–HHS Construction has announced plans to develop a multi-family development of homes in Fort Lee. The Hackensack-based builder plans to break ground on June 1 for a community of starter homes in the north Jersey area.

Make sure the person assigned with the task of media relations is a voracious reader. This person should be reviewing dozens of newspapers and magazines each month to clearly understand the type of coverage the publications offer.

What could have made this story more valuable? Details. How many homes will be constructed in the development? Does the development have a name yet or a specific location? What is the price range of homes in the development? Is any information available yet on the style of homes or amenities offered? Why was Fort Lee selected as the site for the new development?

Although HHS Construction may get some minor publicity by making the announcement as it appears above, they'd certainly gain more if additional details were available that enabled them to create a lead graph that reads as follows:

HACKENSACK, NJ–HHS Construction has announced plans to develop a 150-home development called Riverview in Fort Lee. The award-winning Hackensack-based builder plans to offer colonial-style homes starting at $215,000 in the fully landscaped community. With a location close to New York City and directly on the New Jersey Transit Line, the builder expects high interest in the development, which will feature a community clubhouse, jogging trails on the riverfront, and several swimming pools.

Suddenly the development planned by HHS Construction sounds more appealing and enticing–both to the media and to potential readers of the newspaper story.

For our second example, here are the facts:

- Beacon Designers/Builders plans to have a home in the local Spring Portland Parade of Homes.
- The home was designed internally and is available for sale.
- One of the hallmarks of Beacon Designers/Builders is the use of massive window "sculptures" or "walls of windows" to capture views.

Knowing this information, an appealing way to capture the media's attention might be to write a personal letter to the home editor of the *Portland News*. The Five W's still work in this situation except that we weave them into the letter early on. Here's a look at what the letter might look like:

Stephen Gould
Home Editor
Portland News
125 Bowmont Street
Portland, ME 12345

Dear Mr. Gould,

As you're aware, the Spring Portland Parade of Homes is just weeks away. We'd like to invite you to be one of the first people in the Portland area to have a sneak peak at the enticing home we have designed for this special community event.

Please accept our invitation of a personal tour from our design/build team to see firsthand the magnificent construction elements, detailed interior design work, and breathtaking views offered in our parade of homes entry. One of the hallmarks of Beacon Designers/Builders is the use of massive window "sculptures" or "walls of windows" to capture views of the surrounding area. We're proud that this new home surpasses all of our previous designs in offering panoramic views of the Portland Bay and downtown area.

We know you'll be interested in seeing for yourself the unique design elements we've incorporated into this home. Each room has universal design applications. The home was constructed using low-maintenance "green" building products. And the interior was created to maximize light flow while maintaining energy efficiency for future homeowners.

Sound enticing? It is. Please call me today so that we can set up your personal tour for the week of March 2. We're looking forward to sharing this exciting project with you–and with our fellow Portland residents.

Most sincerely,
Rob Roth
President
Beacon Designers/Builders

The letter above would certainly whet the appetite and interest of any home editor: a personal preview tour, unique design and construction elements, and contact with the top person at the company. This is a win-win situation for Beacon Designers/Builders that could easily result in a full page story in the *Portland News*–all for the investment of a letter and a one-hour personal tour for the editor.

Types of Media and the Unique Deadlines

Interested in communicating directly to consumers? Then you'll want to investigate and develop media lists that target the general public with special interests in home improvement, building, and remodeling. An abundance of media outlets target home buyers, including magazines, television shows, radio shows, newspapers, and websites.

Even though these sound like easy "categories" of media outlets, they each deserve attention and research to perfectly match up your homes with the correct audience. Let's start with consumer magazines. In the building industry, consumer magazines can include everything from project-oriented publications such as *Handy* to product-focused magazines such as *Better Homes and Gardens Home Product Ideas* to regional publications such as *Midwest Living*. Each magazine has a different focus and needs materials specifically created for its editorial needs. Here's a listing of publication categories and samples to consider:

Spotlight on Media Relations 51

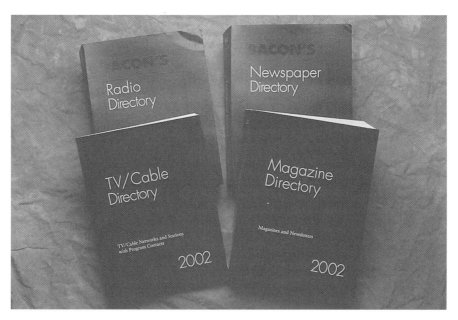

Figure 4-2. The Burrelle's directories can easily help a builder correctly identify the specific media outlets he needs for his press releases and materials.

- Regional consumer magazines (such as *Sunset, Midwest Living, Southern Living*)
- Monthly consumer magazines with a predominantly home and/or gardening focus (such as *Better Homes and Gardens, Home,* and *Country Home*)
- Special interest consumer magazines, issued quarterly, bi-yearly, or yearly, with a special focus on home issues (such as *Better Homes and Gardens Bedrooms and Baths, Woman's Day Home Remodeling,* and *House Beautiful Kitchens and Baths*)
- Home improvement consumer magazines, oriented to projects (such as *Workbench, Family Handyman,* and *Handy*)
- Decorating-oriented consumer magazines (such as *House Beautiful, Windows and Walls,* and *Traditional Home*)
- House plan books (such as *Country Living Dream Home Plans, House Beautiful Home Plans,* and *Better Homes and Gardens Planning Guides*)
- Traditional consumer magazines that have special features or focuses on home improvement (such as *Popular Mechanics, Ladies Home Journal,* and *Martha Stewart Living*)
- Renovation-themed consumer magazines (such as *This Old House, Renovation Style,* and *Better Homes and Gardens Remodeling Ideas*)
- Household project consumer magazines (such as *Better Homes and Gardens Do It Yourself, DIY 101 Decorating Projects,* and *Country Sampler Decorating Ideas*)

Figure 4-3. Consumer magazines can be both general in nature and themed to home improvement. Don't overlook any magazine as a potential showcase for gaining publicity for your homes.

- Special-focus consumer magazines with unique themes (such as *Old House Interiors, Coastal Living,* and *Log Home Design Ideas*)

Just because you're a small- or medium-sized builder doesn't mean that you have nothing to offer some of these large consumer magazines. These publications primarily carry project homes–homes that are submitted to them by proud homeowners, architects, and yes, builders. Having one of your homes featured in a national magazine gives your company additional credibility, exposure in a wide marketplace, and instant recognition. All this, usually at little or no cost to you, the builder!

Although opportunities for builders to get involved with national home improvement television shows are rare, don't ignore broadcast media outlets when developing your consumer media list. You may want to include the business or real estate reporter for local television news stations. Or you may find a local or regionally syndicated home improvement radio show that covers your geographic area. Consumers tuning in to these shows are usually die-hard home improvement fans and are ripe for home purchases.

Getting your media materials to the producers and hosts of these shows should be a top priority for any public relations team. And if your company wishes to get more involved in a radio show, sponsorships and contests are widely available. You can also volunteer to serve as an "Ask the Builder" for live call-in questions from consumers. At no cost to you, your company then becomes positioned as an expert in the marketplace, and potential home buyers get further connected with you and your company.

Individual market research is necessary to pinpoint local home improvement radio shows, but it's easy to locate some of the more popular national shows. Below is a listing of shows you've probably heard of or seen broadcasting live from one of the national industry trade shows:

- *Homefront with Don Zeman*
- *On The House* with the Carey Brothers
- *Ask the Handyman* with Al Haage
- *The Money Pit* with Tom Kraeutler and Mary Baretta

> "Radio is a terrific way to reach home improvement enthusiasts. We've seen a tremendous growth in interest levels over the past several years across the country in radio stations wanting to carry our show. When we started out, we had less than a dozen stations carrying *Homefront*. Now we're available in more than 300 markets in 45 states. That growth speaks dramatically to the effectiveness of radio in reaching targeted audiences."
>
> Don Zeman
> Host
> *Homefront with Don Zeman*

Because magazines and television shows have long lead times in getting information out to consumers, many public relations teams like to add daily and weekly newspapers to their media lists. The potential for more immediate coverage and the ability to target specific geographic markets are seen as top reasons for including newspaper editors on media lists.

As you develop your media list, determine what type of coverage you're looking for and then match it up with the specific editor at the newspaper who writes on that topic. For example, if you're interested in coverage on new developments, make sure the real estate editor is on your media list. Your newspaper's home and business editors should also be on the list and possibly your lifestyle editor if your company is involved in unique projects in the area.

Many daily and weekly newspapers have found economical ways to include home improvement information within their pages without having a staff person dedicated to the topic. Syndicated home improvement writers create a story that is generic in nature (such as how to install a tubular skylight or how to add curb appeal to a home) and supply it to the newspaper. The paper may run a photo with the

> When sending press releases to newspapers, you'll generally receive more coverage if you customize the release to the geographic area. For example, introducing a new show home in the San Francisco market will be more impressive if you include specific information on what features were incorporated into the construction of the home that can best be appreciated by San Francisco residents.

54 *Public Relations for Building Pros*

Figure 4-4. Media directories, like this one on newspaper listings from Burrelle's, are an excellent way to develop a comprehensive and targeted media list for your business.

story or choose to run a weekly column under the writer's by-line. Just as readers know that "Dear Abby" does not write for their newspaper alone, they also recognize that syndicated home improvement writers James Dulley and Michael Walsh write for a variety of newspapers.

In fact, the Dulley column appears in more than 400 newspapers each week and is distributed by Starcott Media Services. The "Home Touch" column by Michael Walsh is distributed by Universal Press Syndicate to over 225 newspapers each week. Popular radio show hosts such as Tim Carter or James and Morris Carey have weekly syndicated home improvement columns distributed in major markets.

Through research you can find the names and addresses of syndicated writers who best match up to your editorial needs and add these individuals to your media list. Offer to serve as resources for these people and you may find your quotes and company name in hundreds of newspapers nationwide!

> Think outside the box. That's what Weiss Development Corp. did when they agreed to create a darkroom in a condominium for a customer in the Chicago area. By expanding the laundry room into an adjacent coat closet and turning an extra linen closet into a coat closet, the Lincolnshire developer helped fulfill a homeowner's dream. More importantly, they shared the story with reporter Pamela Dittmer McKuen from the *Chicago Tribune*.
>
> McKuen later went on to create a front page homes section story on "Living large in a condo" that included reference to the special darkroom. She interviewed other builders who were requested to make unique changes and additions to condos. The trends type story brought lots of publicity to the developers and builders interviewed for the piece.

To really "get your message out there," your company may also try to reach trade professionals such as architects, dealers, real estate agents, and interior designers with your company message. To do this, you'll want to create a customized trade media list.

Just as there are many subcategories to consider for consumer magazines, the same situation exists for trade publications. Hundreds of specialized and generic trade magazines are available by subscription and at the newsstand. Below is a generalized breakdown of the types of publications you can research for your own trade media lists:

- Strictly builder publications (such as *Builder, Big Builder,* and *Professional Builder*)
- Strictly remodeler publications (such as *Remodeling, Qualified Remodeler,* and *Professional Remodeler*)
- Strictly architect publications (such as *Architecture, Architectural Digest,* and *Architectural Record*)
- Specialized industry publications (such as *Residential Architect, Lighting Dimensions,* and *Custom Home*)
- Strictly interior designer publications (such as *Sources & Design, Interior Design,* and *Design Solutions*)
- Strictly dealer publications (such as *ProSales* and *Building Material Dealer*)
- Lap-over construction industry publications (such as *Building Products* and *Design/Build Business*)
- Commercially geared industry publications (such as *Building Design, Commercial Buildings,* and *Multi-Housing News*)
- Market-focused publications (such as *Kitchen and Bath Design News, Automated Builder,* and *Rural Builder*)
- Specialized industry focus publications (such as *Fenestration, Roofing, Metal Home Digest,* and *Contemporary Stone Design*)
- Geographically focused industry publications (such as *Florida Homebuilders, Builder Digest of Northern California,* and *Texas Building Trends*)

The size and circulation of these publications vary greatly. Some magazines, like *Rural Builder,* have only a two-person editorial staff while others, like *Builder,* have over a dozen editors. Knowing and understanding the editorial teams at different publications can help you properly direct your media materials.

At *Rural Builder,* for example, you could send a press release to either (or both) of the magazine's editors and be assured that the entire editorial staff would be saturated with your message. That same strategy is not advisable at *Builder.* With this larger magazine, editors

> With the amazing growth of the Internet as a research tool for consumers, many builders have created websites. The Internet is a terrific way to reach consumers. Make sure your media materials also include your website information so that editors can visit your site for more information about your company.

> "It takes so little time to read through a magazine and really understand the editorial content of the publication. Yet it's amazing how many companies don't do that. They simply 'blanket' the media with press releases. From what I can see, that tactic only results in wasted dollars and effort. Targeting editors with information they can specifically use for their publications is definitely the best strategy for generating product and company publicity."
>
> Paula Doyle
> Former Editor
> *Design/Build Business*

are assigned different beats and topics that they routinely cover. You want to investigate who researches and writes the stories on home design, builder tips, and other specific topics for the magazine and then direct your media materials to the most appropriate person.

Why not saturate all the *Builder* editors with your press release? Because it's wasteful and annoying to editors. It also can be costly, both in the dollars you invest to send out releases and in your company's reputation. Why annoy the new products editor at *Builder* with a press release about a company award that they'll never be able to use for their section of the magazine? Sending such a release only shows that your company hasn't researched the publication's needs. Instead, target your releases carefully to the appropriate person at the magazine. There may be a time when you have news to share with the design trends and business editors that applies to their area of expertise. That's when they'll appreciate receiving materials from you.

Reaching your trade audiences via trade publications is definitely the best route. However, don't forget that these architects, contractors, and designers are also consumers. So the same publications that you target for your consumer media efforts will do double duty for you when they reach building industry professionals.

> Although there are quicker lead times with daily media outlets (like newspapers), a great deal of preparation takes place before a story is run. For instance, a builder may send a press release and photo on a new subdivision to the home and real estate editors of a major daily newspaper such as the *Atlanta Journal and Constitution*. Like most newspapers, the Atlanta paper has a weekly homes section that is planned out weeks in advance. So, even though this is a daily newspaper, it's likely that the subdivision announcement may not get featured for several weeks—if at all—in the publication.

Long lead times are the reality that public relations people must work with to get free press. A company that has no public relations campaign at all and starts from ground zero can expect it to take three to four months before seeing a significant amount of coverage in publications.

Conversely, the media have specific deadlines to meet to continually produce their publications and television shows on a regular basis. Oftentimes this presents an opportunity for you.

Because many editors work on multiple stories simultaneously, it's not unusual for public relations professionals to receive several calls a week requesting materials be

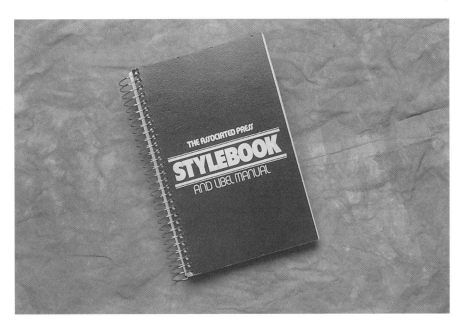

Figure 4-5. Most newspaper and magazine editors adhere to writing standards set forth in the *Associated Press Stylebook and Libel Manual.* Using this manual for your own writing will help your releases be more readily acceptable by the media.

overnighted to media members. Sometimes poor planning on the part of the media is at fault; other times there's an opportunity to fill more pages with copy and photos than previously thought available.

The bottom line is that it doesn't matter. Any request from a media member should be filled as quickly and efficiently as possible. Doing so increases the chances that your company will receive press coverage. The editors will appreciate your fast actions on their behalf and return to your company in the future for more materials.

No Promises

Even though a builder may do everything right, there are no guarantees that he will receive press coverage. For instance, you may send the editor of a major consumer magazine such as *Traditional Home* photos of a house you think is stunning. He calls and requests more information. You promptly comply. And then you wait. The editor may indicate that the magazine plans to run the story in 15 to 18 months or simply wants to have it on hand as filler. Or a newspaper editor may indicate that a story on your home will run on a specific day. Then it gets "bumped out" because the

> If a media person calls requesting information, an interview, or photos for an upcoming story, make certain you clearly understand their deadline. Missing an editor's deadline—even by a day—can often mean getting "bumped" out of the story.

> Remember that even if you are interviewed for a story, there's no guarantee your comments will be used. And oftentimes a 20-minute telephone interview ends up as just three or four sentences in a large story. Most importantly, remember that EVERYTHING you say to a reporter is "on the record" and can be quoted in a story.

paper got more advertisements and needed more space for them.

These are the realities of working with the media. There are no guarantees. Public relations endeavors are vastly different from advertising efforts. With an advertisement, you know exactly the size and look of what will appear. And you pay for it. With public relations, you work hard for your company and homes to be included in stories for free so that readers of the publications—or viewers of TV shows—see the credible nature of your product.

Killing the Myth: Advertising Does NOT Guarantee Story Placement

Have you ever seen it written that "because my company buys lots of ads in a newspaper, we deserve to get lots of free publicity in that paper"? No. It's a myth that should be slaughtered and erased from every person's mind in our industry.

Quite simply, editorial story content and advertising buys are like oil and water. They don't mix. One does not impact the other. Your company may have a six-figure advertising budget for a leading newspaper in your city and never get any press coverage in that paper. Or you may get coverage occasionally because your public relations efforts are solid and match up with the editorial needs of the publication—even if you don't advertise.

Why is this? Because it's the job of the editors to write content for the publication that's of value to their readers. In that sense, editors are "color blind"; they don't care if your company advertises in their publications. As long as you provide timely, accurate, and valuable information that meets their editorial needs, you'll get publicity on their pages.

The bottom line is that no home builder should ever feel they "deserve" to get free publicity in a publication or on a TV or radio show because the company spends money advertising with that same media outlet.

> "Magazines like ours receive many more press releases than we can possibly publish in an issue. While we're interested in gaining tips and information from builders that will help educate homeowners, there's no guarantee that items sent to us will get published."
>
> Rob Fanjoy
> Managing Editor
> *Smart HomeOwner*

Dealing with the Media

Care and Feeding of the Media

When dealing with the media, some people like to sugar coat their messages. That practice won't buy you any friends.

As in most areas in life, honesty is the

best policy. If you're previewing a new home plan to an editor and are asked if the plan comes standard with universal accessibility, think carefully before giving an answer. Wanting to look like the good guy and answer "yes" to the question won't work unless the plan really does comes with standard elements of universal planning–not as upgrades or options but, in this case, as standards.

> Regardless of the amount of dollars your company spends on advertising, your professional public relations efforts is the only thing that will bring you free editorial coverage in a media outlet.

If the answer is "no" but that universal accessibility aspects are optional, be up front and explain that to the editor. Give reasons on why the home is designed the way it is. Bottom line: Never lie to the media.

What happens if you're on the phone with a newspaper editor and you're asked about your company's yearly sales figures? Be honest with the editor. Don't inflate the numbers to make your company look better. And if you are a private company with a policy against releasing sales figures, tell them that. Don't get boxed into a situation in which you're leading the editor on with a comment like, "Well, it's between ten and twelve million dollars."

Consider another situation. If by chance you have used a product that has been recalled or the media has exceptional interest in the product (such as in the EIFS situation), don't try to get around the fact that your company is working on this problem. Speak honestly and openly with the media. You don't need to speculate or guess at answers; only convey information that you know is true.

What to Expect From the Media

Media members need to constantly push information and stories to their audiences. This means they have constant deadlines. They may call you with a request of information and materials *now* and need you to respond quickly to their needs. This is especially true for daily newspaper editors. They need to research, write, and produce stories every single day, and then start all over again in the morning.

Expect the media to be focused and hungry for immediate assistance when they contact you. And because they're so intense in their efforts, they may not have time to visit hundreds of builder homes each year, sit and chat about general topics, or take time to read all the materials you send. You must be ready to respond quickly to the media and be proactive in supplying them with relevant information that perfectly matches their needs.

> For assistance with industry-wide "problem situations" like EIFS, contact your national and local home builders associations to determine what media messages and support are available for you. If you've installed a product that's being recalled, talk to your manufacturer representative about how the company is handling consumer and media inquiries and how they recommend you respond to the media.

> Never lie to members of the media. When you don't know the answer to a media question, tell them you're uncertain. Explain you'd like to do some research and get right back to them. Then do it!

What the Media Expect From You

Media members are generalists; even if they cover a specific beat such as real estate or home improvement they know a little about a lot of topics. They expect you to be the expert about your business and industry and they expect to be able to quote you in their stories.

How can you help media members do their jobs better? By accommodating their requests. Understand that they work on deadlines. Respond quickly to any inquiries. Know how to talk to the media and offer them good spokespeople who can be quoted.

What if media members ask a question that you are uncomfortable answering? One of the first lessons that public relations people learn is that the statement "No Comment" is the kiss of death to media people. Why? Because it can make you look unprepared or guilty of something.

A preferred response to an uncomfortable media question could be, "I'm not sure about that. Let me do some research and get right back to you." Another response could be, "I'm not certain about the exact answer. I'd prefer to give you the most accurate information possible, so let me work on that and call you right back."

Even a tough question that you may have no anticipation of ever answering, such as "How long have you been building homes with unqualified workers and using substandard materials?" is much better answered with something like "I don't believe that's our practice. Let me do some research to get you the most accurate information possible and then I'll get right back to you" than a terse "No comment." Although it's important that you actually DO get back to the reporter with information requested (or a response you prepare with company leaders), it's just as important to "set the stage" for sharing that information.

What if one of your partners makes a presentation at a community function and talks about the growth of your company and expansion of developments in your area? You were not at the presentation, but immediately afterwards you get a call from a local reporter asking for more details. Don't guess or make up information. Tell the reporter you need to discuss this further with your partner and then you'll get back to him. Then do exactly that. Meet with your partner to determine what your

> "I don't expect our public relations contacts to know everything off the top of their heads. It's perfectly acceptable for them to explain they need to do some research or check with someone else for an answer. The critical thing is that I then expect a prompt call back with the promised information."
>
> Colette Ortiz
> Building Editor
> *Home Magazine*

message to the media should be in this case (so that you can reinforce his publicly made statements and appear as a unified team) and then share this information with the editor.

Launching Into Media Relations

When planning to spend dollars on print and broadcast ads, every cent counts. Public relations efforts can be more flexible. Whereas you may have only enough funds to advertise to your top target audiences, it's generally cost effective to reach several primary and secondary audiences with media relations efforts.

Sending out extra press releases and images to media members is an efficient way to get your message out to many target audiences. To run a single print ad in a publication for a secondary target audience may cost $5,000 or more. To tailor a press release to the same secondary audience and send it out with images to a variety of targeted editors could cost less than $500. Thus, in some cases, public relations efforts can be seen more as a "blanketing step" than can the specifically targeted ads that are placed.

As an example, consider Simonton Windows®. This company's key business is manufacturing vinyl replacement windows and patio doors. Their secondary market is manufacturing products for new construction projects. The company's advertising dollars are generally spent in trade print publications such as *Professional Remodeler, Remodeling, Builder,* and *Professional Builder.*

To enhance their overall marketing efforts, Simonton Windows includes media relations strategies in their public relations plans. A continual and aggressive media relations campaign reaches out to editors of remodeling and trade publications with product messages. In addition, the public relations team targets special releases to editors associated with fenestration, manufactured housing, kitchen and bath, interior design, and architectural trade publications. And, although Simonton Windows does not currently advertise directly to consumers, their public relations efforts are also directed to consumer home improvement radio programs, television shows, newspapers, and magazines.

What are the results of this comprehensive media relations effort? Widespread publicity in hundreds of publications each year. Consumers reading tips in local newspapers about when to replace their windows are moved to request more information on product lines from Simonton Windows. Similarly, when the company donated windows to the construction of a custom home in conjunction with the television show *Michael Holigan's Your New House,* consumer

> Customize your press releases for each audience. This means writing specialized releases for completely different readers (i.e., realtors, architects, designers, luxury home buyers, move-up buyers, first-time buyers, etc.) with key messages targeted directly to each audience. An editor is more likely to run a story when it's presented as having a direct benefit for their reader, viewer, or listener.

> Saturate all your primary and secondary media outlets with customized press releases, but do NOT send press releases to publications with readers who don't "fit" with your homes. For example, there's no reason to send *Kitchen and Bath Design News* a press release on how your homes are constructed with low-maintenance exteriors. This publication doesn't cover exterior applications so your effort would be wasted. Worse yet, you'll annoy the editors who will clearly see that you don't understand their magazine or editorial needs.

inquiries skyrocketed. Consumers saw the products on the television show and on the Holigan website and were moved to request more information. This type of pull-through marketing effort supports the overall marketing goals of Simonton Windows. At the same time, it helps develop a broader customer base for the company.

Although builders do not have a physical product like a window to promote, remember that you do have your homes. Creative and aggressive thinking–and planning–can help you promote your homes just as effectively as Simonton Windows promotes its products.

Identify the Right Media for Your Message

Ask yourself this question: Whom do I want buying my homes?

Identifying the right media for your message is as simple as determining what kind of people you want to buy your homes. If you manufacture high-end luxury homes, more than likely you want publicity in high-end city, state, and social publications. Perhaps some of your homes are also targeted to move-up buyers. So you aim your media relations efforts at publications that they read–more newsstand consumer home improvement magazines, daily newspapers, and "influential" target groups such as realtors and community leaders.

One of the best ways to keep in touch with the media is by creating customized media lists for your company. Like an ala carte menu, you can pick and choose the appropriate media lists to use for each of your press releases.

When developing your lists, use media directories available from Bacon's Information, Inc., Media Distribution Services, Burrelle's Information Services, Media Map, or PIMS. Dedicated research is needed to develop and maintain accurate lists.

Developing a media list is not a one-time activity. Consider your list a living and breathing document that should be updated regularly as you learn of editor changes, closing publications, new publications, address corrections, and other changes in the industry. The entire media list also should be reviewed and updated at least once a year.

When developing a newspaper contact list, remember to research specific types of editors. The editor-in-chief of the *Washington Post* won't care at all about a press release he receives concerning the introduction of a new home project in Virginia.

> **"Be familiar with a magazine's purpose and audience."**
>
> Kevin Shoesmith
> **Associate Editor**
> *Workbench*

However, the home editor or real estate editor might be very interested in the same release. Accurately targeting your press release increases the chances that it will get picked up and covered in publications.

To reach radio audiences, developing media contacts takes a bit more effort. First, start with nationally syndicated radio shows dedicated solely to home improvement, such as *Homefront with Don Zeman*, *The Money Pit*, *Ask the Handyman*, and *On the House With The Carey Brothers*. These are weekly shows broadcast simultaneously on anywhere from 30 to 300 radio stations nationwide.

Next on the list would be local area home improvement and real estate radio shows. Some markets, like Philadelphia and Dallas, have locally dedicated programming (usually one hour a week) for home improvement and/or the real estate market. These shows are localized in nature. If you construct homes in Bucks County, Pennsylvania, being a guest on the Philadelphia show *Talking About Your Home* on WPHT with Art McKeown can be tremendously advantageous.

> "Developing and maintaining a valuable media list takes time, research, and effort. I make changes to mine on an almost daily basis to keep it updated."
>
> Kim Drew, APR
> President
> Drew Public Relations

> Who should be on your media list? Most builders are focused on three types of publicity: new projects, show homes/parade of homes, and trends awareness. To gain publicity for your company in these areas, target your media activities to the key editors at the publications on your list involved in real estate, homes, business, and lifestyle.

Planning for Media Attention

Once you've developed a strong media list, it's time to create a game plan. Before sending out materials to the media, read through these generally respected and followed ground rules:

- Don't bombard the media with information not targeted to their audience.
- Space out your media mailings. If you send out several stories at the same time, you're forcing the media to choose which one to run. You may think they're all

> Investing the time and research needed to create a strong media list will lay the groundwork for solid press results. Don't saturate a publication with your materials by sending press releases to every editor at a magazine or newspaper. Select the appropriate editors and target those individuals. This is definitely a case of quality winning, not quantity.

important, but the media won't, and they can't run too many stories for one builder in any issue. Instead, send out releases over a well-spaced timeframe. If you have lots of new project or development information to share, one release every 3 to 4 weeks is appropriate and may get you more coverage than would sending lots of material out at one time.

- Know your publications. Some magazines never run personnel stories, such as the announcement of a new CEO. Don't waste their time (or yours) sending them those types of stories. It's not enough to customize press lists; you need to customize the types of releases you send to publications based on what they need and will use.
- The most concise, well-written releases get the most pickup. There's no two ways about it. Find a person with strong writing skills to tackle press materials and your company will reap the benefits.
- Shorter is better. Press releases are designed like newspaper stories–all the key information should be in the first paragraph. The introductory sentences should give enough pertinent information to fully inform the editor of what the release is about. Media members appreciate short, concise releases that are generally no longer than two pages.

With all this in mind, listed below are several ways to seek media attention.

First, create a personalized "pitch letter" of key information and ideas your company has for a story in a publication. Send that letter to the appropriate editor and see if you get a nibble. What can you offer an editor? A one-on-one interview related to trends in your building market, a personalized tour of new homes, survey results and information from past customers, information shared with you from the manufacturer of your key products on "hot topic" areas like energy efficiency, green and low-maintenance products?

Second, make all your newly targeted media contacts aware of your existence. If your company has not had a public relations effort in the past, it's critical that you let media people know you're now available to assist them. This can be as simple as mailing an introductory letter to the editors on your media lists and including the business cards of your public relations people. Make sure you spell out for the editors exactly what your company does, who your target audiences are for your homes, what makes your company unique in the marketplace, who is available as spokespeople, and what support items you have to offer them (i.e., photography, literature, referrals).

Third, and perhaps most importantly, make sure you have someone designated to handle media requests. If you undertake a media relations campaign and don't follow-up promptly on media requests, your efforts will be wasted. The media should have access to a designated contact person at

> Planning for media attention means being both "softly aggressive" in seeking out publicity opportunities and ready to respond swiftly to media requests. Oftentimes companies pitch story ideas to editors. This works wonderfully when you remember not to be too self-serving in this process and to match up story ideas with editorial content.

your company (or an outside agency) who can assist them at any time. Whether they require a slide image, a person to interview for a story, or a tour of a property, your designated media relations person should be able to assist the media efficiently and swiftly.

Assigning a Spokesperson

It's important to decide who will represent your company to the media. Just because you may be the president of a company doesn't mean you're the ideal spokesperson.

When the editor of *Journal of Light Construction* is doing a story on trends in the roofing industry, he or she may want access to builders using various roofing materials. Perhaps the editor has your company's name from materials you've previously sent them and decides to call. The public relations person should serve as the bridge and the gatekeeper in this situation. He or she should take the initial call and then arrange for a telephone interview with the best company spokesperson on this topic.

Before the phone ever rings, make several determinations that will guide your company in its media relations policies and then communicate those decisions internally and externally. The first rule of thumb is that all media inquiries should be routed through the public relations person.

The second rule is that no one in the company should talk with the media without the counsel of the public relations team. This assures that your company messages are communicated accurately to the media. Getting this message out to all levels of employees–from hourly workers up to the management team–is very important.

Third, determine what type of access the media can have to yourself, your company president, or CEO. Will your top person be available (through your public relations person) for individual interviews? Will this person be available on a limited basis or are you offering widespread availability on all topics?

Making your CEO or president available to the media may get you more coverage and your company may appear completely open and willing to work with the media. But the time commitment needed for this type of media access (including preparation and actual interview time) can be considerable. Small builders may be able to juggle interviews quite easily, but in larger companies access may be more difficult. For this reason, some companies choose to save their top leaders for high-level announcements and company profile interviews.

Finally, every company should

> "Timely responses to interview or material requests is critical when I or one of my writers is working on a story."
>
> Renee Rewiski
> Editor
> *Remodeling News*

66 *Public Relations for Building Pros*

Figure 4-6. "Hot" topics for the media always include low maintenance products. If your company is using innovative products, such as the urethane millwork shown here being installed on a Habitat project, offer to serve as a spokesperson to the local media.

have a distinct list of people available for media interviews. Ideally, an editor would call the public relations person with a topic they wish to discuss with a company representative. The public relations person would then determine who in the company is best qualified—and authorized—to speak to that media person. In addition to your public relations spokesperson, you should consider having the following people on the list:

- President or CEO of the company
- Construction site manager
- Marketing director
- Technical engineer to discuss in-depth technical aspects of products
- Human resources director to handle personnel issues

Evaluate each person individually for their strengths and weaknesses in handling media inquiries. In most cases you should arrange spokesperson training for these people so that they know specifically how to deal with the media over the telephone, in person, on the radio, and on television.

People selected to serve as media spokespersons should be fully aware of their responsibilities, which include being available to the public relations team at all times and providing quotable information to the media. These people should be well versed on your company's key target messages. They should also know what company information is confidential and should not be shared with media representatives.

Also keep in mind that you may wish to select entirely different spokespeople for print versus broadcast interviews. With telephone or in-person print interviews, a person's knowledge level is the most important factor. With radio and broadcast interviews, knowledge levels are critical but the person must have a strong energy level, appear confident and comfortable, be well spoken, and represent the company well.

> There is no right or wrong number of spokespeople a company can have. The most important thing is to prepare anyone who will talk to the media beforehand. This means getting everyone to agree on key messages because ANYTHING they say during an interview can be used by an editor in print.

Media spokesperson training can be a great help to many company leaders. You may be the most knowledgeable builder in the world, but if you stutter when the camera starts rolling, your company will look unprofessional. Some people gravitate toward media interviews, others are extremely self-conscious and hide from them. Invest time and training to find and train the spokespeople who will work best for your company.

Do's and Don't's of Working with the Media

Media members have a job to do: They need to research and report on stories. You can make their lives easier and potentially enhance the coverage your company receives by keeping the following guidelines in mind when dealing with them:

Do:
- Take time to listen to media inquiries, think, and then respond.
- Promptly return all media calls and fulfill media requests.
- Know your company's key messages and incorporate them into discussions with media members.
- Ask questions, both to clarify questions and to find out more about the story they're working on.
- Follow the guidelines of *The Associated Press Stylebook and Libel Manual* when writing materials for the media.

Do Not:
- Answer any questions before you are fully prepared to do so; don't feel rushed. Make sure you feel comfortable with an answer before giving it.
- Say "No comment" to anything. It sounds too terse, as if you have something to hide. Instead, offer a statement such as "We're not prepared to discuss that today. Let's move on to something else and we can always revisit that issue at another time."
- Make negative comments about your competitors.
- Lie or make up information.

Ever believe that anything you say to a media person is off the record. If you don't want to read about it or see it on TV, don't say it.

With all these thoughts in mind, it's sometimes helpful to have some guidance on handling media questions–whether in person, during a recorded interview, or on the telephone. Consider these tips:

- *Relax . . . and smile*
- *Stick to the script.* Remember your key messages and keep weaving them into your remarks.
- *Keep your answers short and complete* during television questions, . Answer in full sentences, but don't run on and on.
- *Don't guess.* If you don't know an answer, or think there's someone better qualified to handle the question, tell that to the reporter. You can always hook them up with another person at the company.
- *Talk in simple language.* You're the expert. Make the media person understand your answer by using common terms, not technical jargon or industry-specific language.
- *Speak naturally and clearly.* Stop from time to time to ask the media person if they understand what you're saying and if you're on the right track with providing the information they need.
- *Don't fidget.* Even during a telephone interview, your full attention should be given to the media person, not to other items on your desk.
- *Give your business card to the media person* or make sure to spell your name for them. Also make sure that they have your title and the company name correct.

For television interviews, there is another set of guidelines to keep in mind. In addition to the items listed above, here are some tips for preparing for a television interview:

Clothing
- Wear comfortable, familiar clothes suited to the environment or situation. Business suits don't work when you're doing a walk-through of a house under construction. Ask the producer what attire he suggests you wear.
- Medium to dark solid colors work best on camera.
- Never wear solid white shirts, bold patterns, checks, strong stripes, or pinstripes for a television interview.
- Tuck in shirts smoothly and pull up pants at the waist before the interview starts.
- Select low-wrinkle clothing.

Movement:
- Stand erect, but not stiff. Try to be natural.
- Don't slouch or lean up against anything.
- Don't sway, twist, rock, or lean.
- When sitting, lean slightly forward to show interest.

- Never fold your arms over your chest in a protective gesture.
- Don't fall into the "waving arms" routine. Using arm and hand motions is fine, but you don't want to be flailing around.
- Don't fiddle or fidget with pencils, jewelry, paper, etc.

Eye contact:
- Maintain direct eye contact with the interviewer, not with the camera, unless directed otherwise.
- Don't get distracted during the interview by activity around you. Keep focused on the interviewer and the questions.
- Relax . . . and smile.

Searching for still more tips on meeting the needs of the media? Several media members list the following as their pet peeves:

- Never staple slides directly to press releases; they're difficult to get off. Paperclips are also difficult because they can puncture the slide image. The best scenario is to attach a slide sleeve to the release and insert the slide into the protective plastic sheet.
- If you're sending print photography, do not paperclip it to a press release. After a few days in the mail, the paperclips crease the photo.
- Label slides and photographs whenever possible with the company name and descriptive information, especially if you are submitting multiple images simultaneously. Phone numbers on the slides are also helpful.
- When submitting a CD or zip disk, include a printout of thumbnail images and file names. Many editors working on PCs cannot read the CDs.
- Try to avoid including people in photographs whenever possible. Installation shots showing subcontractors working are appropriate, but Vanna-like models showing off a refrigerator typically end up in the trash.
- Include only contact names on press releases of the people you want called. Don't try to impress the media by putting the president's name on the release if he or she is not willing, available, and eager to handle media calls on a regular basis.
- Courtesy counts. If you're calling to pitch a story, ask if it's a good time for you to discuss an idea with them.
- Do not overload media representatives with follow-up calls. A company representative who calls too often becomes a nuisance. When that happens editors tend to think negatively about the company and its products.
- Know that your press release paper and format–plus spelling and punctuation–reflect the standards of your company.

"Use professional photographers—digital cameras are not the answer."

Naomi Anderson
Publisher
Sources & Design

> "While editors try to work as far in advance as possible, often we are juggling the research and writing of several stories simultaneously and need information and images very quickly. The public relations person that sends a photo within a day or two gets my endless thanks for making my job easier."
>
> Katy Tomasulo
> Managing Editor
> *ProSales*

Send out only materials you would be proud to show your top customers.

"Faster-Than-A-Speeding-Bullet" Responses

Want to snag a quote in a newspaper or magazine story related to the homes your company builds? Eager to see yourself interviewed on a local television station? Searching for a way get photos of your homes in local newspaper stories? Sometimes it's as easy as returning a phone call–quickly.

As mentioned earlier, speed is critical when dealing with the media. Media members work on deadlines and they need builders to respond promptly to requests for materials so that they can meet those deadlines.

Although not every media call requires a split-second response, you'll develop a solid reputation for bending over backwards to work with the media if you treat every media call as urgent. Why let a media person's call go unanswered for five or six hours? The only possible result is that the frustrated reporter will go to another source–most likely your competitor–for a quote or for information.

Out of all the pink While You Were Out messages that cross your desk each day, responding to your company president and any media person should always be tops on the list. Media members who need to arrange an interview with you, get a photo, or talk about industry trends will quickly turn to your competitor if they don't receive immediate assistance from you. And once your competitor makes a positive impression on a media member, that's a media contact you've lost.

Making Your Story Newsworthy

There's an old journalism saying that goes something like this: If a dog bites a man, it's not newsworthy. If a man bites a dog, it's front page news. So it goes for builders trying to get publicity.

> Make it a priority at your company that media requests are handled promptly and efficiently. When you do this, editors will learn they can depend on you and they'll repeatedly return to your company as their primary source for materials.

Generally, builders should understand that the introduction of a new home or the expansion of a subdivision may not always be earth-shattering news. That is, unless you *make* it so. How do you do that? By having your public relations person create a press release that has flair, excitement, and insight.

First, let's consider a paint company that

is introducing its 57th shade of white interior paint. How can such a potentially mundane product introduction be made newsworthy? The company can add some research elements that explain why this particular shade of white is being introduced. They can provide details on how and where the shade of white will work best in the home. They can quote credible sources—such as interior designers or builders—in their press release to explain the differences between white shades and what sets this one apart. Or they could give the product a name that demands press attention, such as Electric White or Blizzard Ice. Simply reporting on the new color may not be newsworthy, but adding a unique spin on the product could make it more enticing to editors . . . and consumers.

As a second example, let's tie in to a typical home builder. Joe Builder

Figure 4-7. When contractors for Heivilin Remodeling took on a renovation project in Virginia, they quickly found themselves the focal point of stories in *Professional Remodeler* magazine. The unique project brought the company both print and broadcast media attention.

constructs starter homes throughout the Dallas area. His company already has five developments that have fairly cookie-cutter designs, and now Joe is introducing the exact same design in a new development within a mile of the older development. Sounds boring, right?

This is where a creative public relations person can have some fun. Why is Joe building so many homes in that area and with that style? Is there a story angle that suggests massive growth in the area, that Joe hit on a terrific location that consumers are demanding, or that his homes are so popular he's being "forced"

> "A widget is a widget is a widget. Nothing new, and nothing newsworthy. Until some smart public relations person paints it green and shows how it helps save the environment. Then it's news and it gets in my magazine."
>
> Joyce Powell
> Former Editor
> *Building Components*

by home buyers to construct more homes? Is one of the homes going to be a milestone home–for example, the 500th home he constructs? Could he hold a raffle for one of the homes and donate the funds to a charity? What does Joe think is the secret of his success, and is he willing to share the secret with editors?

An angle can always be found. It just takes digging and creativity to find it time after time and then get media coverage.

Looking for more ways to make a story newsworthy and increase your chances of getting published? Try the following:

- Include documented research showing how your home(s) fills a niche.
- Incorporate third-party quotes from outside sources (such as interior designers, realtors, architects, previous home buyers, builder association leaders) offering stories of how your homes fill a specific need in the community and how these people feel about your construction skills. Don't be too gushy. Endorsement quotes must be realistic and balanced.
- Give complete details about your projects and your company. Sounds easy, but it's amazing how many press releases are written in such a fluffy manner that they leave out the dimensions and critical features of the product. Those are important details for editors, so fill them in!
- Be a visionary. Have your company offer insight into the industry and make suggestions as to what will happen in the future. No one expects you to have a crystal ball, just to offer expert opinions. Look at the December and January issues of trade magazines and you'll quickly discover how many trends stories "get ink."

Taking Advantage of Current Trends

There's a general rule that most editors are willing to share–to gain publicity, you must have newsworthy materials. The trick is that what is considered newsworthy by one publication may be of no interest to another publication.

As an example, if your company hires a new CEO, that's worthy of a press release announcement to your local newspapers and to select trade publications in the industry that carry personnel announcements. However, it would be impossible to find a consumer home improvement magazine, television or radio show interested in covering it. If you're a large builder, the CEO announcement may be of interest to newspapers near your headquarters location, the new CEO's hometown, or even his college alma mater. Beyond that scope, however, there may be little or no interest in your company's new leadership.

The best way to assure your company of ongoing and positive publicity is to focus on newsworthy items such as major new subdivision and development introductions, trends research announcements, case study profiles, participation in show and parade of homes programs, and award-winning projects. Although you should still write separately for different types of publications (e.g., consumer versus trade magazines), you'll generally find greater acceptance and interest with these types of press releases.

Another great way to gain publicity for your company is to ride the wave of current and popular trends. It's easy to identify trends in the building industry—energy efficiency, use of environmentally sensitive products, installation of products offering safety and security for homeowners, and products geared to low maintenance have all been hot topics for the past decade. More than likely these same categories will continue to offer opportunities for stories into the future. And as baby boomers grow older, products that ease them into the retirement lifestyle, including those focused on accessibility, will gain in popularity.

Figure 4-8. Have you recently built a home with lots of windows and doors that allow light and air to flow through a home? If so, consider creating a story for your local media on the connection homeowners are seeking between their interior and exterior living space. Any home has an angle for a story . . . it's the job of the public relations professional to find that angle and maximize it. (Courtesy of Crossville Porcelain Stone/USA).

What other topics are currently getting play in the media? Open any magazine or newspaper and you're sure to find a story related to mold and mildew problems. Several years ago it was the EIFS situation. The energy crisis still gets media attention several times a year, as does the issue of questionable construction quality and the need for more skilled laborers.

> If a construction or home industry topic suddenly gets hot in your area, move swiftly and decide your company's position on the issue. Then call up local reporters and offer your thoughts or have your public relations team issue a press release on the topic.

To jump on this bandwagon, determine where your company "sits" on several of these issues and then promote your position. Certainly it would help if your company had a 10-point checklist to reduce mold and mildew problems with top-flight construction steps. But you can also benefit if your company installs products with features directly related to current trends, such as laminated safety glass in windows or vinyl siding or even top-of-the-line insulation for maximizing energy efficiency. All of these elements give you a wonderful opportunity for spinning your story several different ways and maximizing publicity opportunities.

Once you have determined your position on these issues, offer to serve as a resource on the topic for local reporters, provide viewpoint papers to the media, or present free homeowner workshops. Invite reporters to attend any workshops you give and you may get free press on the session. Or include a written recap on your website and then in post-session press releases distributed to the media.

> Invest the time to know and understand your target media and then tailor different press releases and mailings to each one to increase your chance of gaining widespread publicity.

What if your company is not on the cutting edge of using products associated with new trends? Then use your creativity as your company develops and starts new trends and patterns of its own.

Equally important as determining what newsworthy and trendy materials you should send the media is recognizing what types of press releases should *not* be sent. Materials that are too self serving will not be used by most editors. With minimal exception, here is a listing of topics that most media people in both trade and consumer arenas would not cover in their magazines:

> "The amount of press releases that end up in my trash is unbelievable. I have to ask myself, 'Do these people even read my magazine?' If they did, they'd know we never cover personnel announcements, safety award presentations, and annual sales projections. Those are just pieces of paper completely wasted on us."
>
> Peter Whiteley
> Senior Writer
> *Sunset*

- Individual home price increases or decreases
- Announcements of position changes below the company leadership level
- Quarterly earnings and loss statements for your company
- The introduction of strictly promotional programs geared to potential customers
- Annual reports

How do you determine what a publication would consider newsworthy? Read.

Make it a priority to consistently read the key trade and consumer publications on your media target list.

Leads On Unique Projects

Both consumer and trade media appreciate when builders share unique projects with them. Magazines like *Better Homes and Gardens* and *Coastal Living* have teams of magazine scouts set up in cities across the country. Their job is to research and find unique projects for their magazines to cover. If your company adopts the mindset of a magazine scout, you can enhance the opportunity to work more closely with magazines and some television shows.

There are several ways to inform media members of unique projects. In some instances you may wish to publish a case study sheet on a project and offer the photos and story to any publication interested in covering it. Public relations people for builders may also create written listings that describe projects (much like a real estate listing) and distribute them to media members to gauge their interest. An "adopt-a-project" type of sheet can offer brief information to editors and recommend that they contact you for more details.

Many magazines, such as *Luxury Home Builder* and *Southern Living*, require exclusivity for the stories they run. This means they will not carry a pictorial story on a project if another magazine has covered the same project. Patience and good media contacts is the name of this game.

If you're interested in offering a project exclusively to a magazine editor, send them a pitch letter with initial photos of the project. Explain why the project is ideally suited for their publication, the scope of the project, and that you're offering the project exclusively to the magazine. Set a time limit for a decision by the

---

The Home Builders Association of Metro Denver certainly knows how to stay on top of trends. Several years ago the organization created the industry's first green building checklist for individual homes. As a follow-up, the Colorado HBA has rolled out the first large-scale energy and resource efficiency guide for community development.

According to a story in *Professional Builder* magazine, the Built Green Communities program is a cutting edge program created to assure that entire communities are "built green." The comprehensive concept encourages developers and builders to think about the entire planned use of the land with green construction in mind.

Publicity throughout Colorado for the program—and the builders involved—has been very strong. Given the cutting edge development of the Built Green Communities concept, it's also generating outstanding national consumer and trade publicity for everyone involved.

When pitching an exclusive project to a magazine, be sure to obtain a written agreement from the editors stating that your company will be showcased in either the story itself or in a photo caption—and also ask to be listed in the resource guide at the end of the story or in the back of the magazine.

editor and then follow up. If the editor accepts the project, you've achieved a wonderful opportunity to have a project featuring your home showcased in the magazine. If the magazine rejects the story, try again with another targeted magazine.

Positioning Your Company as Experts: How to Get Exposure When There Is No Hard News

It happens to every company and every public relations person periodically: The well of new home introductions simply dries up. Does that mean you can't generate ongoing publicity for your company? No. Follow the old adage, "When life gives you lemons, make lemonade."

Creativity comes into play when there is no hard news to share with the media. As discussed earlier in this chapter, a skilled public relations person can supplement truly new stories with trends stories, by-lined articles, and hot topic stories. Another way to assure that your company name stays top-of-mind with editors and your target audiences is to position your company (and its management team) as leaders in the field.

Assess the strengths of your company spokespeople. You'll most likely find you have people on your team with installation knowledge, industry insights, and in leadership positions of various organizations. Promote the availability of your people to serve as experts for phone interviews with editors, as guests on radio shows, and as installation wizards on television home improvement shows.

Not sure where to start? What product do you most want publicity for? Systematically review its features, benefits, marketplace challenges, and where it "fits" with its competitors. As an example, let's consider a model home available in the Chicago area–the Rosemont subdivision created by Gordon Builders. The furnished model, called Sonnyside, is now a year old. Sales are slow but steady. Tod Gordon, the builder, decides to pull out all the public relations stops he can to generate more attention for the model home. Working with his public relations team, he develops a plan that calls for:

1. Offering the model home to small community groups as a meeting location.
2. Making the model home available to the local realtors association for a meeting and offering to include refreshments.
3. Analyzing the unique features and benefits of the home and then promoting these via press releases to targeted local media.

"We're constantly searching for top-quality residential coastal projects to highlight in our magazine. We consider it a 'win-win' situation when a company or builder supplies us with an exclusive opportunity that includes professional photography and lots of information on a coastal home."

Cathy Still Johnson
Home Editor
Coastal Living

4. Working with a local charity and radio station, he uses the theme "help make this house a home" and a special weekend is designated for a community open house. Consumers are invited to tour the home and fill its cabinets with food, towels, clothes, and toys that will be donated to the charity.
5. Hosting a series of move-up buyer seminars in the model home over several weekends.
6. Contacting the leading manufacturers of products in the home to see if their company public relations people can help promote the model home.
7. Offering the home for a live remote setting for the *Wake Up Chicago* show for one week.
8. Sending a postcard to past home buyers offering them something for free (or cash) when they bring a friend to tour the model home on specific dates.
9. Hosting a reception at the model home with information on the subdivision for corporate relocation personnel in the surrounding area.
10. Throwing a one-year anniversary party for the Rosemont subdivision, celebrating the success of the community and promoting the event to the media.

A different type of building industry situation clearly shows that companies can gain publicity even when no new product introductions are involved.

In 2001, Therma-Tru Doors sponsored an independent research study involving more than 2,000 consumers nationwide. The online survey requested that qualified

---

"Hire a sharp public relations person. They make it easy to request information, press kits, and/or photography. Also, invest in beautiful photography. We make good-quality shots bigger, and readers really respond."

Sarah Wolf
Associate Editor
Better Homes and Gardens
Do It Yourself

Every day of the year consumers are scratching their heads, looking for answers to questions. One Atlanta television station, WSB-TV (an ABC affiliate), created a unique way to help the public get answers. WSB-TV offers a service called "Wsbtv.com EXPERTS" on their station website. Consumers can visit the site 24 hours a day, locate one of dozens of topic areas, and ask an expert for advice. Categories include home, home improvement, and construction. And savvy builders eager to position themselves with the public are offering their services as experts on the site.

As an example, when a visitor with a roofing question clicks on that topic, a full page of information on the local company "Dr. Roof" pops up. There's information on the company, a direct link to their website, and a list of frequently asked questions and answers. The visitor then has the opportunity to ask personal questions that will be answered within 24 hours.

This is one way you could promote your business in your area. If your local radio or television stations don't offer the service on their website, suggest that they do so! For more information on the Atlanta site, visit www.wsbtv.com.

homeowners review images of home exteriors and estimate how much each home would cost. The groups were shown unenhanced images and then the same homes upgraded with the door manufacturer's fiberglass entry systems, including decorative glass, sidelights, and transoms.

Therma-Tru shared the results with the media and received strong publicity on the study. The findings, which suggested a grand entrance can significantly increase the perceived value of a home, played out very well for the company. One industry publication, *Window & Door* magazine, ran a two-page story on the survey results, including quotes from Therma-Tru executives and the images shown on the survey.

By sponsoring and sharing this research study, Therma-Tru literally made hard news for their company. At the same time, they positioned themselves as an industry leader and gained outstanding press coverage for the company.

This public relations program worked very well for Therma-Tru. A savvy builder who uses Therma-Tru products could easily run with this same information (supplied by Therma-Tru) and promote the same topics to local media. Then, the builder can serve as the resource for the papers to interview, attesting to the fact that entranceways really do make a difference in homes and offering tips to homeowners. One company's solid campaign can be a springboard for a builder's follow-up efforts on a local basis.

Chapter 5

Winning at Community Relations

Like many companies, yours is community oriented. More than likely you're located in a small- to medium-sized city or town. Many folks know your work and your company. When it comes time to sponsor Little League teams, advertise in the high school yearbook, and participate in local charity fundraisers, your company does its share. But are you leveraging that participation? Is your company really involved in worthwhile activities that best reflect your organization's core values? Are you participating in community programs that match up with the profile of your target home buyers?

In community relations activities, volunteerism, and charitable giving, many builders generously share their dollars, time, and employees. The key to successfully doing so is recognizing that these are important marketing opportunities that should be maximized for the good of everyone involved.

The Perfect Fit: Getting Involved in Projects and Programs that Make Sense for Your Company

There are a host of reasons why builders and their companies should get involved in community projects. By becoming more involved in your local community, your company appears civic-minded and friendly. This strengthens your image as a good builder and corporate citizen.

The good–and interesting–thing about community relations is that your company's donations of time and resources generally are paid back to you in a variety of ways. When your company allows a manager extra time off once a month to participate in school board meetings, this reflects well on the company. So does a parade float sponsored in the town's Fourth of July celebration, dollars or time donated to build a new library, and a company-sponsored clean-up day in the city park.

Will these efforts help you sell more homes? Definitely. Maybe not directly from a specific person, but overall the impact is strong and people are impressed. Why? Because you are positively positioning your company's image while enhancing your reputation, something that's needed if you are going to be top-of-mind when people are ready to buy or build a home.

As a builder, it makes the most sense for you and your company to participate in community activities focused on construction activities. For instance, your company may wish to participate in the construction of a playground at a

> Your company does not have to carry the burden of supporting every community project. Pick and choose those projects that you feel strongly about *and* that will benefit your company.

local park or a local housing project in conjunction with Habitat For Humanity or another worthy organization.

Although there may be hundreds of community projects that could use your company's support, try to focus on ones that work for you. The local senior citizen center may need volunteers to help organize afternoon activities, but it might be much more productive for your company to offer to help build an addition to a local homeless shelter.

Charitable Giving Versus Volunteerism: Finding the Perfect Balance for Your Company

Look around your community. It won't take long to identify programs and projects that need your support. There is an abundance of worthwhile organizations that can use volunteer man hours, dollars, and product donations. Consider women and family shelters, orphanages, hospitals, low-income housing projects, or local beautification programs.

Search for opportunities to link up with projects that support the housing industry or relate to construction in some way. Tie in with a local parade of homes project benefiting a charity. Support a mass build project in your community. Or arrange for your company to supply sheltered workshop employees with special projects. The ideas and the list are endless.

A popular program supported by builders worldwide is Habitat For Humanity International. Founded in 1987, this non-profit organization helps create affordable housing locations. Many builders have already found great benefits from working with either the overall international organization or with local Habitat chapters on individual and mass project builds. By involving your company in an organization like Habitat For Humanity, you're keeping a community and nationwide focus on the industry your company supports.

Another popular building project is Rebuilding Together. This is the nation's largest volunteer home rehabilitation organization in the country. Over a million volunteers provide support to low-income families with children, to the disabled, and to elderly people. This organization has more than 250 affiliates and welcomes support from companies of all sizes.

> "The key to making community relations involvement successful for your company is to maximize the potential of your involvement. There's always the rewarding feeling you'll get from supporting your community. However, with proper coordination and attention, there can be so much more to gain for your business."
>
> **Chris Monroe**
> **Vice President of Marketing**
> **Simonton Windows**

Figure 5-1. On a job site at the 2002 International Builders' Show, builders and manufacturer's representatives volunteered to construct five homes for Habitat for Humanity that were later moved to different locations in Atlanta.

Looking for more projects that are both national and local in scope? Consider Ronald McDonald House Charities. The organization has more than 174 local charities serving 32 countries. In addition, the popular Ronald McDonald Houses are built and run worldwide to help children and families in need.

One of the more important aspects of giving deals with heart versus money. You've heard the saying "Actions speak louder than words." When it comes to involvement in the community, actions can often speak louder than dollars.

Simply because your company employs people and is a part of a local community, more than likely you will get requests for support of local projects, organizations, and fundraisers–everything from food drives for homeless people to sponsoring youth soccer teams to pledging dollars for employees participating in walk-a-thons.

One of the most effective ways to maintain control of these donations is to have a system in place for handling donation requests–and then communicate this system to all employees. Whether you decide to have the function outlined and handled by the human resources department (or the public relations team) or create an employee-organized task force to make determinations, make certain that key management members set the overriding guidelines.

> Make sure to maximize your company's involvement in any charitable organizations–especially when donated products or services are involved.

> Searching for a way to help? Contact your local NAHB chapter to find out what programs it's working on through the Home Builders Care Initiative. Hundreds of projects have been started by chapters and members nationwide.
>
> If you're in the Seattle area, you may have heard about Seattle's Master Builders Association and their work with HomeAid to build the "Vision House" shelter. The multifamily transitional housing complex provides 20 private bedrooms and group gathering areas for homeless women and children.
>
> A similar structure was built by the HBA of Northern California to offer a safe haven for hundreds of families. The East County Family Transitional Center was constructed on about an acre of land and consists of three two-story structures housing 20 apartments.
>
> On a larger scale, home builders nationwide showed they truly cared after the horrific terrorist attacks on September 11, 2001. Thousands of members of the housing industry contributed more than $9.8 million to the Home Builders Care Victims' Relief Fund. NAHB provided $250,000 in seed money for the fund, supplemented by $125,000 by the National Housing Endowment. From there, collection of monies grew, from $25 to over $1 million from builders impacted by the tragic events of September 11.
>
> Led by caring leaders of the housing industry, Home Builders Care is a worthwhile organization that can involve builders at all levels through chapters nationwide. To learn more about the efforts underway, visit www.homebuilderscare.org.

Consider these steps before giving money and volunteer time to support local opportunities:

- Make sure that all company giving matches your mission statement or stated company goals. For instance, if your company is dedicated to family values, it makes the most sense for the charities you support to have direct tie-ins with helping families.
- Create a set yearly cash budget for company donations (such as ads and sponsorships). Designate additional resources and limits for donations of paid employee time to worthwhile projects.
- Determine and distribute a set policy for employee solicitations during company work time (and on company property) for everything from selling Girl Scout cookies to participating in raffles and sweepstakes.
- Decide and communicate if your company will offer matching funds to employees for cash donations to non-profit organizations.
- Consider how "dead stock" products or damaged products will be handled at your company. Will these be donated to organizations? Who makes the decision?
- Determine if your company will allow employees to use any of their paid time (whether accumulated work, sick, or vacation time) to volunteer their services to community organizations.

Checklist of Local Opportunities

If your company chooses to get involved in a national charitable effort, make certain you have the staffing and resources available to support such an endeavor. To maximize the benefits of a national opportunity, you need solid marketing, sales, and public relations support. There are hundreds of worthy organizations that could use your support. When your company is ready to consider a commitment to a national charitable program, these groups are available and eager to talk with you:

- Habitat For Humanity International–912-924-6935
- American Red Cross Chapters Nationwide Surplus Product Program–314-727-8900
- Ronald McDonald House Charities–630-623-7048
- Rebuilding Together–800-473-4229
- SkillsUSA VICA–703-737-0603

> There are occasions when it's more productive for your company to align itself with a smaller charitable organization than with a larger one. Groups like Habitat For Humanity have a high profile and are worthwhile, but your company may be one of hundreds participating and get lost in the shuffle. By sponsoring and supporting a smaller organization you can serve as the lead player and generate massive amounts of publicity for your efforts.

Looking for more ideas for national community and charity projects? Visit *Worth* magazine's website at www.worth.com for a listing of "America's 100 Best Charities." The organizations are broken down into the categories of relief and development, health, education, environment, and human resources. The article also gives guidance on how to choose the charity that may work best for you along with listing organizations that should be avoided.

Figure 5-2. Whether your company volunteers manpower for a national or local project, the important factor is to link in with projects that support the housing industry or relate to construction in some way.

Figure 5-3. Contact your local chamber of commerce, United Way or American Red Cross chapters to determine what construction projects in your area can benefit best from your company's involvement. Before committing to a project, ask to speak with the organization's public relations person to determine how your efforts will be promoted in the marketplace.

> Call your local home builders association for community relations project ideas. Oftentimes they're a clearing house for organizations requesting assistance from local builders for special community projects.

Your company may also wish to explore the website www.give.org sponsored by the Better Business Bureau Wise Giving Alliance. The Alliance reports on nationally soliciting charitable organizations that are the subject of donor inquiries and has a special section for businesses considering making donations. Reports are available on more than 250 charities, along with tips for giving, charity standards, and a wise giving guide published quarterly by the Alliance.

Need ideas for easy, local projects? Check with your local chamber of commerce to see if there are any projects planned that could use some extra manpower. Ask your employees if their families are involved in any community programs that could use your support. Contact your United Way or American Red Cross office to inquire about local projects that need assistance.

Leading the Way: Creating a Solid Community Image

Why? That may be a question you're still asking yourself. Why should I invest my time, talents, energies—and sometimes dollars—in community relations projects?

Aside from the fact that participation in community programs can be personally rewarding for you and your employees, having your company involved visibly and actively in community projects creates a solid company image.

This image is important when building and selling homes. If you're meeting with homeowners and find out they have children, wouldn't it be great to tell them that your company helped build the local playground? Or that your company sponsors and coaches the youth softball team?

Let your good deeds and actions lead the way to setting your image a notch above the competition in your market. Focus on selecting and getting involved in visible and valuable community projects that set you apart and you'll be doing your company a service. And remember that every action counts. If you participate in your daughter's PTA and volunteer to help paint the school's media center on a Saturday, you are building community relations. You'll be networking with other parents–your potential customers–and building relations.

> One of the keys is to focus your volunteerism efforts on projects that bring exposure and credibility with target markets to your business. If you're a custom home builder, don't volunteer to assist with low-income housing projects. Instead, volunteer to build a tennis court in a community park or landscape the property at a church. Match up your efforts with people who may be of value to your company in the future.

> Giving a speech to a community group? Make sure you send a press release to the local media beforehand. They may be interested in attending and covering the speech for their newspaper or broadcast.

The Importance of Local Home Shows

Have you visited your local home show in the past several years? You may be surprised to hear that many builders answer "no"–that they're either too busy or it doesn't interest them. If that was your answer, think again. Why? Because local home shows bring out community citizens, who are potential home buyers.

If you're interested in covering all the bases to reach potential customers, you should know and understand who attends local home shows. Many times it's people looking for new ideas, new products, remodeling options, or those who are just frustrated with where they're living now. That's the perfect audience for a home builder. Even if a person is just strolling the show looking for ideas, you want to be the builder they touch base with, the one who will stick out in their minds when they are ready to buy a new home.

To connect with this important group of consumers, start with your research.

> If you're a larger builder with the funds available, consider sponsoring scholarship programs in your community. That's what Florida builder/developer WCI of Bonita Springs did recently. They jump-started a scholarship program for low-income families with a $5 million donation to FloridaChild.
>
> According to a story in *Builder* magazine on WCI's donation (which also got them extensive publicity), FloridaChild is a Tampa-based organization that promotes education initiatives. Most states have non-profit groups like this one and can be found through your statewide school systems office.

Determine what home shows are held in your area, when they're held, who sponsors them, and how other builders participate. Look at this as an opportunity to reach out to community residents. Remember to also determine how much booth space costs and if your local suppliers can assist you with co-op funds for the space, displays, literature, or other marketing and merchandising materials.

Next, talk with the show coordinators. Tell them you really want your participation in the show to stand out. Suggest any of the following to maximize your involvement:

- If the show has an official radio, TV, or newspaper sponsor, offer to serve as a professional resource. You can be on a radio show and have listeners call in to "ask the builder" a question. Or you can offer a column of tips to a newspaper or TV reporter to run before the home show so that you can promote your involvement. Acting as a resource to consumers will get you attention.
- Offer to construct a child's fantasy playhouse for the home show. Have it in your booth or at another prominent location on the show floor with signage that the playhouse is provided by your company. Use a creative sign like, "Fine home construction is child's play to us at ABC Builders" or "If we can do this with a playhouse, imagine what we can build for you!" Suggest the playhouse be raffled off to benefit a charity or donated afterwards to a local playground.

Finally, look inside your own exhibit booth to see how you can really sell your company. Try any of these ideas:

- Include large photos of the homes you've constructed.
- Display endorsement letters and quotes from past customers.
- Send a pre-show postcard to potential and previous customers inviting them to stop by your booth.
- Have experts on hand to answer home construction and financing questions.
- Invite local realtors to help you staff your booth.
- Incorporate models and miniature displays into your booth.
- Push your best home features. Whether it's sunrooms, expansive decks, or master suites, show your "best stuff" to the public.
- Work with suppliers of your favorite products to incorporate their displays into your booth.
- Offer prize drawings from your booth to build up your follow-up list.

If you're unfamiliar with working a trade show on a regular basis, here are some fast and easy tips for you and your team:

> Don't be discouraged if you find out that other builders don't participate widely in home shows in your area. They're missing out. This gives you an open door to make an impact on the home show attendees while strengthening your company's community visibility.

- Have everyone in the same "uniform" in the show booth—khaki pants and your company's logo shirt.
- Everyone should wear nametags.
- Never sit down. Don't even allow chairs in your booth. You look much more approachable and friendly when walking around your booth.
- Don't eat or drink in the booth.
- Schedule breaks for booth workers so they have time to walk the show and eat and drink outside the booth.
- Keep the booth neat and tidy–just like you would your jobsite.
- Restock literature and business cards regularly, and offer them to booth visitors.
- Warmly approach visitors. Don't be too pushy or standoffish. Find the approach that works best for you.
- Have a trash container in or very near your booth for people to deposit their trash rather than leaving it on your displays.
- Encourage your booth workers to spread out in the booth–don't huddle together or block the entranceway to the booth or you will seem unapproachable.

Maximizing Community Relations Efforts

Your company can maximize its involvement in community relations projects in a variety of ways. When it's time for you to get involved in a local activity, consider the following:

- Look for ways to enhance your company's visibility at a project location. This may mean offering to construct an "in progress" sign at the site of a local playground build that includes your company logo.
- Turn your employees into walking billboards and supply them with logo T-shirts to wear while doing community project work.
- If your company is being asked for services such as paying for printing up raffle tickets for a group, agree to do so only if your company's logo is included on the ticket.
- Make yourself or a company spokesperson available for local radio show interviews on community projects to explain how your company is involved in supporting the community and this specific project.
- Notify local newspapers in advance when your employees and/or key management team members will be working on a community project. Don't

> Don't be afraid to issue local press releases on your company's involvement in community projects. Although some people may think the act is self-serving, it's more important that good deeds in the community be shared with others. Who knows? Your company's enthusiastic participation in an activity may spur another company to contribute their time and energy in the community.

forget to issue a local press release. Even if the community organization is handling media relations, remember to support their efforts by issuing your own releases.
- Remember to supply community organizations with background information on your company and logos for inclusion in their press releases and annual reports.

Sharing the Good News

Just as with successful media relations efforts, it's important to share the results of community relations, charitable giving, and volunteerism activities both internally and externally. If you're a larger company with many employees, consider promoting your company's involvement in non-profit efforts in the following ways:

- Dedicate a story in each issue of your employee newsletter to community relations success stories.
- Develop a slogan or name for your company's volunteer efforts to rally people. Hyatt Hotels has the "Hyatt F.O.R.C.E . . . A **F**amily **O**f **R**esponsible, **C**aring **E**mployees."
- Post photos of employees volunteering for projects on company bulletin boards in break and time clock areas.
- Salute individual employees whose spirit of volunteerism far surpasses normal activity levels with a special yearly award or recognition during employee functions.
- Include information on your company's volunteerism and donation policies in literature for new employees and during new hire training sessions.
- If, and when, your company receives recognition for its participation in a project, share the recognition internally. Make sure any awards or plaques are rotated in different employee areas so that everyone feels like a winner.
- Hang a banner near the employee entrance congratulating workers for making a difference in their community.

> During your company's annual picnic, Christmas party, or pancake breakfast, make sure to recognize and offer a round of applause to all employees who volunteered their time for community projects during the year. This type of recognition impacts how employees feel about your company.

Externally you can promote your efforts in the following ways:
- Send out press releases to local, regional, and national media to announce milestones in your company's community relations efforts.
- Proudly include information about your company's community relations efforts in your annual report and on your website.

- Discreetly include information on your company's volunteerism efforts in sales literature.
- If your company receives external recognition for its participation in a project, share the recognition with the local media via a press release.

> Nominate your company's shining star volunteers for local volunteerism awards or, better yet, have your company sponsor a recognition program for the entire community.

Chapter 6

Partnering for Success

Remember the old saying "No man is an island"? Truer words could not be spoken in the housing industry. There are thousands of home builders throughout the world trying to figure out the best sales strategies to use, the most effective public relations tactics to employ, and the most impactful marketing ideas to implement. Why not join efforts?

Although it's important that builders develop their own marketing and public relations plans, it can also be beneficial to partner up with others in the industry from time to time to get the most out of your marketing and public relations efforts. Whether it's working with several builders to promote a community parade of homes project or working with product manufacturers on promotions, partnerships between non-competitive companies can spell good news for everyone involved.

Looking for a good example of teamwork that is breeding success? The Wood Promotion Network got together in 2001 to create the "Be Constructive" promotion. Within months and through aggressive public relations and marketing efforts, the program conveyed the benefits of framing houses with wood. The group–made up of manufacturers associated with the wood products industry and builders using wood products–launched a website, staffed a 1–800 answer line, and gave dozens of media interviews. As a result, the "Be Constructive" tagline is getting lots of positive play in the press and quickly becoming an easily recognizable slogan in our industry.

Another unique example of partnering for success is the promotion being conducted by Target Stores. Lindal Cedar Homes in Seattle, Washington, and architect Michael Graves have agreed to design and build a house for the grand prize in a sweepstakes for newlyweds sponsored by Target Stores' Club Wedd. The partnership has netted nationwide publicity for everyone involved, including trade and consumer stories on the builder, the architect, Target Stores, and the promotion itself.

Hitch a Ride–Working Through and With Local Home Builder Associations

As any builder knows, there are plenty of associations available in the building industry. Generally, most reputable builders are members of the National Association of Home Builders (NAHB) and their local home building associations. There's also a variety of special interest organizations, including everything from the National Roofing Contractors Association to the National Sash & Door Jobbers Association.

Depending on the level of interest your company has in the building products it uses, opportunities abound for getting involved with specialty organizations such as the Gypsum Association, the Structural Board Association, or the Home Automation Association.

To maximize your company's membership investment in an association, it's important to do more than just pay dues and read the monthly newsletter. Serving on committees and attending educational seminars allows you to network with other industry representatives and promote your business. More importantly, getting to know the marketing, public relations, and public affairs staff members for the associations can lead to great public relations opportunities for your company.

Industry association contacts can help your company in many ways. Consider these ideas:

- Link your company's website to associations you're connected with to provide a service to your website visitors and add extra credibility to your own site.
- Provide association website people with stories, photos, and materials to enhance their site (and reflect positively on your company).
- Sponsor scholarships and educational industry programs

Eager to find out which associations might be a good match for your company? Glance through this listing and visit the websites of these leading industry associations for more information. You'll find both general and specialized associations that can complement your company's public relations activities. As an added bonus, many associations may match up perfectly with your target audiences. Membership in these organizations can get you closer to your customers!

American Architectural Manufacturers Association–*www.aamanet.org*
American Fence Association–www.americanfenceassoc.org
American Forest & Paper Association–www.afandpa.org
American Gas Association–www.aga.com
American Hardboard Association–www.hardboard.org
American Hardware Manufacturers Association–www.ahma.org
American Institute of Architects–www.AIA.org
American Institute of Building Design–www.aibd.org
American Institute of Constructors–www.aicnet.org
American Lighting Association–www.americanlightingassoc.com
American Society of Home Inspectors–www.ashi.com
American Society of Interior Designers–www.asidnews.com
American Solar Energy Society–www.ases.org
American Subcontractors Association–www.asaonline.com
American Wood Council–www.awc.org
America Wood-Preservers' Association–www.awpa.com

APA—The Engineered Wood Association—www.apawood.org
Appalachian Hardwood Manufacturers, Inc.—www.appalachianwood.org
Architectural Woodwork Institute—www.awinet.org
Asphalt Roofing Manufacturers Association—www.asphaltroofing.org
Association for Safe and Accessible Products—www.asapdc.aol.com
Association of Construction Inspectors—www.iami.org/aci.html
Association of Home Appliance Manufacturers—www.aham.org
Association of the Wall and Ceiling Industries Intl.—www.awci.org
Bath Enclosure Manufacturers Association—www.bathenclosures.org
Brick Institute of America—www.bia.org
Builders Hardware Manufacturer Association—www.buildershardware.com
California Redwood Association—www.calredwood.org
Canadian Home Builders' Association—www.buildermanual.com
Carpet and Rug Institute—www.carpet-rug.com
Cedar Shake & Shingle Bureau—www.cedarbureau.org
Ceiling & Interior Systems Construction Association—www.cisca.org
Ceramic Tile Network—www.tilenet.com
Composite Panel Association—www.pbmdf.com
Copper Development Association—www.copper.org
EIFS Industry Members Association—www.eifsfacts.com
Environmental Information Association—www.eiacom.com
Florida Building Material Association—www.fbma.org
Forest Products Society—www.forestprod.org

through associations your company is affiliated with to strengthen the industry.

- Work with association public relations people to maximize your visibility to media people they're working with on stories.
- Make certain industry association public relations people know who they can refer media inquiries to at your company. Share your company's key messages with these same public relations people so that when they receive media calls they can intelligently speak about your company and entice the editors to speak with you.
- Offer to provide company spokespeople for association educational and media sessions to position your company as an industry leader.

Use these same ideas when dealing with your local builder association. Remember, you want to stand out in the marketplace in all possible ways. If there's a Home-A-Rama planned, get involved. If the association is supplying tips on maintaining the home to the local media, offer some comments or to serve as spokesperson. If there's a library that needs building, volunteer your workers to assist in the construction. Be visible and build awareness for your efforts!

Joint Builder Efforts

Looking to share the burden and the success of some larger marketing efforts? Look to other builders in your area that you respect. While you probably don't want to join efforts on too many projects, there may be occasions when builders–even competitive ones–can join efforts and everyone gains.

Consider a community playground that needs to be constructed. If four builders join together to construct the playground, everyone benefits. A sign at the construction site credits all builders involved and your company only needs to contribute one-quarter of the workforce.

Another worthwhile community effort might be for several builders to form a summer internship program for students. Five builders could design a five-week course for senior high school students. Each week the students rotate to a different builder–and each week they are exposed to a different construction skill. Rather than teaching students about framing, window installation, drywalling, millwork and tiling, your company may just focus on the framing aspect.

Looking for another joint builder effort? If several builders are constructing homes in a single subdivision, this is the ideal scenario for a shared special event. Consider coordinating efforts so that each builder has an open model home on a specific weekend and then throw your own parade of homes. Enhance the event by hosting a joint pool party or tennis tournament to generate interest in the subdivision's amenities. Or really make a hit with

Gas Appliance Manufacturers Association–www.gamanet.org
Gypsum Association–www.gypsum.org
Hardwood Council–www.hardwoodcouncil.com
Hardwood Manufacturers Association–www.hardwood.org
Hardwood Plywood & Veneer Association–www.hpva.org
Hearth Products Association–www.hearthassociation.org
Home Automation Association–www.homeautomation.org
International Wood Products Association–www.iwpawood.org
Kitchen Cabinet Manufacturers Association–www.kcma.org
Louisiana Building Material Dealers Association–www.lbmda.org
Lumbermen's Association of Texas–www.lat.org
Manufactured Housing Institute–www.mfghome.org
Metal Construction Association–www.mcai.org
National Association of Floor Covering Distributors–www.nafcd.com
National Association of Home Builders–www.nahb.com
National Association of the Remodeling Industry–www.nari.org
National Fenestration Rating Council–www.nfrc.org
National Hardwood Lumber Association–www.natlhardwood.org
National Housewares Manufacturers Association–www.housewares.org
National Kitchen & Bath Association–www.nkba.org
National Lumber & Building Material Dealers Association–www.dealer.org
National Oak Flooring Manufacturers Association–www.nofma.org

National Paint and Coatings Association—www.paint.org
National Roofing Contractors Association—www.nrca.net
National Sash & Door Jobbers Association—www.nsdja.com
National Tile Roofing Manufacturers Association—www.ntrma.org
National Wood Flooring Association—www.woodfloors.org
National Wooden Pallet & Container Association—www.nwpca.com
North American Insulation Manufacturers Association—www.naima.org
North American Wholesale Lumber Association—www.lumber.org
Northeastern Retail Lumber Association—www.nrla.org
Northwestern Lumber Association—www.nlassn.org
Paint & Decorating Retailers Association—www.pdra.org
Plumbing Manufacturers Institute—www.pmihome.org
Portland Cement Association—www.portcement.org
Power Tool Institute—www.taol.com/pti
Screen Manufacturers Association—104200.266@compuserve.com
Southeastern Lumber Manufacturers Association—www.slma.org
Southern Building Material Association—www.southernbuilder.com
Southern Forest Products Association—www.sfpa.org
Southern Pine Council—www.southernpine.com
Structural Board Association—www.sba-osb.com
Structural Insulated Panel Association—www.sips.org
United American Contractors Association—www.engineers.org
Western Forestry & Conservation Association—www.teleport.com/~Ewfca

parents by jointly funding a babysitting service at the subdivision playground for the afternoon while parents tour your homes!

Promoting With Manufacturers

A unique public relations strategy is to join efforts for a project with building product manufacturers. This type of activity is limited only by your own creativity. The results can be especially appealing for all parties. You can share expenses and work on a project that benefits everyone involved.

Here are some ideas for builders to consider when working with building product manufacturers:

- Find out if local building product manufacturers plan to host media tours of their facilities during the year. If they are, offer to serve as a guest speaker to give the media a clear view of the building projects in your state. This positions you as a leader in the industry and gets you connected with more media contacts.
- Make manufacturers aware of when you're using their products in especially visible projects, like show homes and parade of homes projects. Manufacturers are eager to find installed settings with good interior design. Oftentimes they will pay for photography (or offer to split the cost with you) of the home and allow you to use the images for your own promotional purposes.

- Co-sponsor consumer focus groups with manufacturers on several common denominator topics of interest to both of you. This way each company gains valuable research at a fraction of the cost of sponsoring the research independently.
- Select a company with products you regularly use and believe in. Offer to co-host deskside briefings with the manufacturer in key editor cities; they talk about their product and you talk about using it to enhance your homes.
- Ask manufacturers about their model and show home

Western Red Cedar Lumber Association–www.cofi.org/wrcla
Western Wood Preservers Institute–www.wwpinstitute.org
Western Wood Products Association–www.wwpa.org
Window & Door Manufacturers Association–www.wdma.com
Wood Floor Covering Association–www.wfca.org
Wood Molding & Millwork Producers Association–www.wmmpa.com
Wood Truss Council of America–www.woodtruss.com
World Forestry Institute–www.vpm.com/wfi

Figure 6-1. Looking for a unique way to gain publicity? Create a collection of playhouses and then raffle them off for charity at a local mall or state fair. (Courtesy of Style Solutions).

> Divide and conquer. That's the theme for association memberships. Have your technical people participate in more technical associations and your marketing team members involved in more marketing-oriented organizations of subcommittees of associations. Maximize your involvement in associations by having everyone at your company get involved.

"For several years in the 1990s a group of four or five builders who belonged to our local home builder association got together and created a village of children's playhouses. There was everything from a café to a bank to a grocery story.

We had them on display in the local mall and raffled them off for charities. We approached Georgia-Pacific and they gave us many of the materials for the playhouses. Many of our subcontractors did volunteer work on the playhouses. Overall, it was a terrific public relations program for the mall, for the charity Voices for Children, and for each builder. Most importantly, our participation helped us give something back to our community and kept us in front of our key target audiences."

Bob Goodier
President
Goodier Builders

programs. You may find appealing incentives for including their products in your homes.

- Make photos, written copies, and a link to your company's website available to manufacturers' websites in exchange for doing the same for them. This raises the opportunity for several companies to reach similar target audiences. You'll be surprised at how many people visiting a window manufacturer site will be interested in linking to your site to see projects using those same windows.
- Offer to serve as a referral to a manufacturer when they need builders for media stories. Oftentimes a manufacturer gets a call from a trade magazine asking for the name and number of a builder using their product. You want to be the builder they refer to so that you can gain more publicity for your company and homes.
- For large show home projects, determine if there are any cost savings in hiring a shared photographer with another company to capture photos of the homes and products in use.
- One builder, Dobson Construction, created a beachfront retreat using dozens of Simonton windows. The public relations team at the vinyl window manufacturer was able to place the story exclusively with *Luxury Home Builder* magazine. This resulted in great publicity–an eight-page cover story–for both the builder and the manufacturer.

Working with Dealers

Extend your public relations "arms and legs" to dealers and distributors you regularly purchase products from. These are people you are probably in touch with on an almost daily basis, so they should be able to assist you with promotional opportunities.

More than likely you'll have to educate your dealer contacts on the value of working on joint public relations activities and on your company's overall goals and objectives. Once you've done this, there are endless ways to get dealers to work with you.

An excellent way to increase awareness of your company is to work with dealers who participate in local home and garden shows. Consider working trade show booths with them, having your brag book of project homes in their booth, or supplying company literature for them to distribute. Why would they do this? To show the public that they are associated with quality builders and to help you grow your business.

If your dealer or distributor also services the general public, request that they support your selling efforts by distributing flyers promoting your special parade of homes project or open house. They may be able to do bag stuffers, hang your banners, or co-sponsor a direct mail postcard (that includes your company's information) that is sent to their homeowner customers.

> Ask your manufacturer's representative if discounted (or free) product is available to you for use in model homes or a local parade of homes in exchange for allowing them to photograph the project and distribute literature during the open house tours. Work closely with manufacturers to pitch truly unique projects to magazines.

> When builder Bill Akers was tapped to construct the 2000 *Southern Living* Idea House® in Tennessee, he worked closely with Weather Shield® Windows & Doors. Akers, president of Akers Custom Homes of Tennessee, used the window manufacturer's new Legacy Series™ line of windows in the magazine's show home.
>
> According to his remarks in a Weather Shield publication after the event, "I've never had a show home before that generated so much attention for the windows. People were drawn to them. They'd walk through the house and reach out to touch the different windows, turn the cranks, and feel the sturdiness." For Akers, the project home not only brought his company outstanding publicity but also connected him with a manufacturer whose products are a perfect fit for the five to seven custom homes he builds each year.

"There's no reason for builders to give each other the cold shoulder. Sometimes the best way to get recognized is to unite—join efforts and conquer the marketplace together!"

Scott Grote
Owner
Grote Construction

Some dealers hold "Ask the Builder" sessions in which people can come into the dealership on a scheduled Saturday or weekday evening and ask any questions they have. Again, your visibility and prestige increases with your target audiences when you participate in a program like this.

98 *Public Relations for Building Pros*

Figure 6-2. This home, built by Dobson Construction, was featured as a cover story in the August 2002 issue of *Luxury Home Builder*. Public relations people at Simonton Windows® worked to "place" the story in the magazine—and also created a four-page case study brochure on the home.

Figure 6-3. Many manufacturers create case study brochures showcasing unique projects that use their products. Connecting with manufacturers on your homes can result in free publicity and sales pieces for your company. (Courtesy of Simonton Windows®)

If your company has decided to become involved in a community relations project such as constructing a set of bleachers at a local soccer field or rebuilding a library, talk to your dealer. He can go back to his suppliers/manufacturers and see if product can be made available either for free or at deeply discounted prices.

Remember to also ask your dealer about programs available from different manufacturers. Model home programs, parade of homes, and community programs are all offered by many manufacturers who want to gain awareness and usage of their products in your area. The key to successfully working with dealers is to brainstorm as many ideas as possible and leave no stone unturned when it comes to seeing what manufacturers can do for you and your business.

Special Magazine and TV Show Projects

Wouldn't you like to have your new home project featured on the cover of *Home Magazine?* Or have Home and Garden Television call and request an interview? That's the private dream of most builders—to gain widespread recognition for projects.

> Offer to host new homeowner information classes at your dealer's location. They do the work of promoting the sessions and inviting people (which helps them increase their visibility) and you do the presentation, answer questions, and network with potential home buyers.

> Want a terrific example of builders working together for the good of their community? In Atlanta, Georgia, there is a consortium of 10 large contractors. Formed 13 years ago, the group is called GlenCastle Constructors. Their sole reason for existence is to focus on projects that will improve their communities.
>
> What was supposed to be a one-time project of converting an old city prison into 68 apartments for homeless people has turned into an ongoing union. The Atlanta Day Shelter for Women & Children was rebuilt by the group. A shelter was added for new mothers at a midtown Atlanta synagogue, a 300-bed camp was constructed for sick and disabled children near Atlanta, and a warehouse was turned into a shelter. Nine projects have been completed, including a warehouse for the Atlanta Food Bank.
>
> In a story in the *Atlanta Journal and Constitution* newspaper in March of 2002, Norman Koplan, director of the Atlanta Bureau of Buildings, remarked "I don't know of any other city that has a group of construction executives like this." The normally competitive builders share the work in each project and coordinate donations of suppliers, design experts, and volunteer groups for each project.
>
> Need advise on how to coordinate such an enterprising group in your community? Contact one of the following GlenCastle Constructors members: Pinkerton & Laws, Humphries and Company, Batson-Cook, Beers Construction, H. J. Russell, Hardin Construction, Holder Construction, Malone Construction, or Flagler and Winter Construction.

Unfortunately, opportunities like this don't just fall in your lap. You have to make them happen. And, while not all builders are going to get the call from a magazine editor or television show producer, some

Figure 6-4. Home improvement television shows, like the ones shown here, are always eager to learn about unique home remodeling and new construction projects. Having professional photography taken of your home to "pitch" to television show producers is the first step in getting on the air.

builders *do* get the call. But usually only after they've done their homework and made contact with key media people.

First you need to evaluate your company and projects. Determine which ones match up with the look of specific magazines or television shows. If you truly feel you have a project that is exceptional and would look great on the pages of a particular magazine or featured on a television show, here's how to proceed:

- Match the style and components of your project with a magazine or television show offering the same style. For example, if you construct log homes, consider approaching the editors of *Log Home Living*. If you build high-end, multi-million dollar custom homes, target your projects to the editors of *Custom Home* magazine.

> "We appreciate it when builders request our products for their model homes. Use of our products in model homes is a great opportunity to support the builder and it gives Alcoa Building Products an excellent way to showcase our products. We see it as a 'win-win' situation for the trade and for us."
>
> Thomas G. Maher
> Director of Marketing
> Communications and Programs
> Alcoa Building Products, Inc.

Partnering for Success **101**

Figure 6-5. Building product manufacturers, like Weather Shield Windows and Doors, oftentimes include builder projects in their newsletters, customer communications materials and website.

- Take professional pictures. You'll need these to share with the editors. If you're only in the planning stages, get copies of the blueprints or artist's renderings–anything that can be visually shared with the media.
- Contact the home or special projects editor at the magazine or the senior producer at the television show. Explain–in writing and with visual backup–the project you're proposing that they cover.
- If the project has not yet been constructed, discuss with your media contacts how you can involve them in the construction steps (e.g., allowing access for film crews, working with building product manufacturers they may like, etc.). If the project is already finished, include a

> If you're approached by a magazine or television show to serve as the builder of a "home of the year" or other special project, know that this requires a great deal of time and effort on your part. Due to photo shoots and filming, your construction schedule may be extended past what you're used to, and you may be required to use certain products from specific manufacturers. Usually these products are offered for free or at deep discounts. Although this can be an incentive, it also means working with many companies on multiple delivery and installation schedules.

> "For our home of the year projects, we look for builders who have a proven track record of producing outstanding homes. These builders need to be flexible to work with, accommodating of our schedules, and open to new products and ideas."
>
> Joan McCloskey
> Editorial Director
> *Better Homes and Gardens*

Getting Former Customers to Promote Your Company

Imagine this situation: You're at a wooded lot in a subdivision meeting with potential customers, Mr. and Mrs. Clarke. You've been discussing the options of building a home for them on the site. They seem interested but not overly enthusiastic.

Just then, Janet Collins, the homeowner down the street, drives by. Mrs. Collins stops and rolls down her window. She leans out to say hello to you. Then she looks directly at the Clarkes. She gives them a big smile and tells them how they should "snatch up" your services quickly because you built her family the very best home they've ever lived in. How's that for an endorsement?

Although you probably won't be fortunate enough to get the Janet Clarke's of the world to "drop in" on all your potential customer meetings, wouldn't it be great if you could? That's where keeping

letter from the homeowners authorizing the magazine or show to feature the house and giving them access for their own means.

Magazines and television shows working on special projects will need to take their own photography and film. Your professional images are just to sell them on your project. Much time and effort goes into working with magazines and shows, so you may want to have a public relations professional tackle this challenging project for you.

What's the best of all worlds for a builder? Being sought out by a major consumer magazine to construct their home of the year project—and getting paid for it! This type of situation allows a local or regional custom builder to gain generally unprecedented press coverage with the national magazine and with local media outlets.

Builder Ralph Haskins, owner of Haskins Homes in Waukee, Iowa, discovered the flexibility needed by his team when selected to construct the 2001 Home of the Year for *Better Homes and Gardens* magazine. According to an interview he gave to an internal Weather Shield Windows & Doors publication (another source of publicity for him), "The term 'construction by group decision' is a good way to define the building process when dealing with a magazine." He continued, "One of the best aspects of the project was that the magazine editors introduced us to a variety of products I had never dealt with previously."

Haskins worked closely with several manufacturers, including Weather Shield, in the construction and promotion of his house. The window company supported Haskin's efforts by creating case study profiles on the home and promoting Haskins nationwide to their vast audiences.

Partnering for Success **103**

Figure 6-6. Get it in writing. If you've created the perfect home for a customer, don't hesitate to request a letter from them singing your praises. Also, offer to send a photographer to the home once they're moved in, so that you both have professional images of your showcase home to share with future potential customers. (Photo courtesy of Style Solutions)

former satisfied customers happy can really pay off for the future success of your company.

The power of positive word of mouth isn't a myth. It can sway people in the purchase of many things. Thus the dedication you put into building your company's reputation is just as important—before, during, and after the sale of your homes—as the building practices you use when constructing your homes.

Need ideas to on how your past customers can help promote your company? Try any of the following:
- If you have a website or company newsletter available for external audi-

> It's important to create good relationships with your customers. As many people know, an unhappy customer will tell far more people about your company than will a happy one. So minimize any negative attention from the start by focusing on customer satisfaction.

> While writing this book, I personally purchased a new home. It was just two and a half years old and the current family was being relocated from Georgia to Kansas. They were sick about leaving such a great house. Once I saw the home, I could understand why. The floor plan was terrific and the construction appeared solid. The home

inspector could not find one settling crack inside or out, and gave it an A+ rating.

Just three weeks after moving in, I was out walking. My exuberant Golden Retriever pulled me right up to a gentleman across the street who had just gotten out of his truck. After a friendly rub for the dog, he inquired if I liked the neighborhood. Long story short, I told him I had just moved here and loved it. I proudly pointed out my new home—and he just as proudly informed me that he had built it!

What did I do? I started thanking him profusely! I rattled on and on about how great the house is, the glowing inspector's report, and how much I was enjoying it. We had some more friendly conversation and then parted ways.

Certainly the builder must have felt good about our encounter—I did. But once inside the house I thought about it and realized he really missed a great opportunity. If he had been on the ball and thinking about his business, he could have offered me one of his business cards and suggested that any other friends of mine thinking of building a home should call him. Since I was such a happy customer, that would have made sense.

He also should have made some attempt to find out who I was. More importantly, he missed out on a strong opportunity to have a second-time home buyer endorse his company. Had he asked, I would have been happy to write him a letter detailing all the great things I said about his product—and he could have used that to share with future customers. Or, had he asked, I would have eagerly chatted with any of his potential customers considering the same floor plan. As I'm still in the first three months of homeownership, he could have also found a way to drop off a "welcome to the neighborhood" card in my mailbox. That would have strengthened my feelings about his company. He did none of these things, and missed all these opportunities.

The moral of the story for any builder? Don't ence viewing, offer a rotating column of "thanks and congratulations" to people who have purchased homes from you.

- Feature one family on your website or in your newsletter in a profile story every quarter. Make them feel like celebrities. Include a few comments they have made about your home (e.g., how much they're enjoying living there, how spacious the layout is, how the home is a perfect fit for their family, etc.) that will be appealing for prospective customers to read.

- Send a brief survey/questionnaire to past customers about a month or so after the sale of a new home. Inquire about general things and also about specific recommendations they would have for your business. Ask them to write in any comments they have about your company. When you find really great comments, share them with new customers and as endorsements in your literature.

- Keep a "brag book" of letters you get from satisfied customers in the lobby of your office and sales area for potential customers to review at their leisure.

- Host an "ABC Builder Appreciation Party" for all the people who have purchased homes from you during the past year. Make this casual event a thank you to past customers and an opportunity to renew contact with them. Consider allowing them to bring friends or family with them. Who knows? They may know people ready to build a new home with you!

let even a chance encounter with a satisfied customer go by. Be mentally ready to take action. At the very least, exchange contact information (a business card for them and their name and address for you) so that you can build your database of satisfied customers.

"When I met the builder of our home, I felt like I was meeting a celebrity. The house was relatively new, but we weren't the original home buyers. We saw the wear and tear the home had been through, yet it was still in good structural shape. It was obvious the builder had invested a lot of himself in the construction and I respected him for that. Now I proudly show off the home as 'ours' and heartily recommend the builder to my friends!"

Genie Gordon
Homeowner

Chapter 7

Employee Relations

Without exception, every builder has at least one employee. And once you have just a single employee, you need to consider how you communicate information. As it's more likely that most builders have dozens—and sometimes hundreds—of employees, developing a strategy for sharing information becomes critical.

Some companies consider employee relations a part of the human resources or personnel function. That's fine as long as the public relations and marketing team can be an integral part of developing and enacting the employee relations programs. Why? Because the marketing division chiefly develops the key messages for a company and communicates them to outside target audiences. It's important that communicating these same messages consistently to internal audiences be given equal consideration and attention.

Sharing the Vision and the Message

One of the most important things a company can do is share its vision with employees at all levels. Whether it's a prominently displayed mission statement framed and posted in the main office, a letter sent to all employees at home, or an annual report supplied to employees, it's vital that employees understand where a company is going—and how they can help the company achieve success.

Consider a football team. There's no way a coach would throw all the players on the field, point at the goalposts, and say "go win the game." There are play books to learn, strategies to develop and implement, team camaraderie to build. That's what it should be like within every builder's company.

Determine who the "coach" is at your business. Usually the role falls to the president, owner, or CEO. Create the game plan and then share the pertinent ideas with all levels of employees—even with your key subcontractors. By bringing employees on board with the plan you're empowering them to be part of the company's success. Having one company leader to rally around also provides employees with a focus and a clearer understanding that they're part of a larger team effort.

> Determine the primary and secondary languages of all your employees and subcontractors. You may find out that employee communications pieces need to be developed in several languages to gain widespread understanding amongst your worker base.

Figure 7-1. Make certain that every management team member, employee and subcontractor know and understand the vision for your company. Only when everyone knows of your deep commitment to your business can they embrace the same level of dedication. (Photo courtesy of Style Solutions)

A construction worker who distinctly understands the goal of creating top quality homes with zero defects knows what his role is in the company. Suddenly, quality control becomes not just a company catchword but a way of life for your company and your workers.

Getting Your Team to Promote the Company

By informing employees of your company's goals and direction, you're doing more than educating employees–you're creating an ever-increasing resource of communications people.

Consider how many people outside your company that a single house framer or painter speaks with in a week. There are family members, social friends, church members,

"There's no such thing as communicating too much with your employees. The more they know about the company's direction and goals, the more they become committed to being part of the growth of the company."

Bruce Johnson
President
White River™ Hardwoods/
Woodworks, Inc.

> "Employee communications is a vital part of our company. I can't imagine not sharing our vision and goals with employees at all levels at ODL. We take steps to make certain our employees are well informed about day-to-day activities at our plants and our long-term mission for success."
>
> Dave Killoran
> President
> ODL, Inc.

fellow club members, sports buddies, and other folks. One employee could talk with perhaps 75 to 100 people in a week.

First, let's take a worst-case scenario. One of your employees reports to work and is responsible for laying concrete foundations, day after day, at your developments. The company doesn't share goals and information with him, so all he knows is that his job is boring and redundant. What type of discussions do you think this employee has with his friends over a mug of beer at the end of a day?

Now consider the same employee in a different situation. In this scenario he's sent monthly company newsletters with stories on the company's direction and successes. During bi-weekly meetings his supervisor motivates the employees toward quality construction practices and explains why their roles are so important to the success of the company. A quarterly report is mailed to his home highlighting different parts of the construction team and their vital role in the overall success of the company.

This employee has a greater understanding of his role in the company and where the company is headed in the future. Over a mug of beer, this employee is more likely to say positive things about the organization. And this employee may be more committed to staying employed at your company than switching to a different builder for an hourly increase of a quarter.

Once your company takes a step in this positive direction, it's just another short leap to actually motivating employees at all levels to promote the company. In this case, consider the word "promote" loosely. A satisfied employee may help a company recruit other employees. There's also the hope that team members will share your company's growth message with the community, thus improving your company's image.

It's as important to develop and communicate company messages to the management team as it is to hourly employees. As many management team members are in constant contact with outside professionals, customers, and sales contacts, every member of the team should be fully aware of the company's vision and goals.

Tools of Employee Relations

Empowering employees should not be a once-a-year pep rally. Companies that are truly successful weave their messages into everyday company communications to employees–they reiterate their mission statement and company goals on a continual basis.

Curious as to how this can be done? Consider using the following communications tools at your company to share information with employees:

- Bulletin board postings
- Paycheck stuffers
- Break room banners
- Jobsite posters/banners
- Signage near time clocks
- Flyers stuffed under windshield wipers in your parking lot
- Direct mail materials to employees' homes
- Employee newsletters
- Customized website specifically for employees
- Mass e-mails to employees
- Quarterly letters to employees notifying them of company status
- Pre-shift meetings
- Employee barbecues, picnics, breakfasts, shift parties
- Videocassette, audiocassette, CD, or DVD motivational talks from company leaders

Figure 7-2. Wondering how to reach your construction team with your company messages? Try paycheck stuffers, direct mailed literature to their homes, employee newsletters and worker incentive programs. (Photo courtesy of Style Solutions)

> Consider using these communications tools to keep employees informed: shift meetings with supervisors, bulletin board messages, break area posters, and paycheck stuffers.

- Closed circuit television reports from company leaders in break areas
- Publishing annual reports
- Creating worker incentive programs

With such an array of communications tools available there's certainly a mixture that can work well for any builder and his company. The key is not which tools are used. The key is using a variety of tools consistently to constantly impact your employees with your company message.

Sharing Positive Results

Positive feedback and information are great motivators for employees at all levels. The results of a successful public relations campaign can be exciting and invigorating to share with employees.

As your business gains favorable press clips, make certain to share these with employees. Employees will soon begin to see the impact your public relations program has in the marketplace, which should lead them to taking even greater pride in their company.

Consider the following ways of sharing positive publicity and public relations results with your employees:

Figure 7-3. Share publicity results with your employees by posting clips on bulletin boards, including them with paychecks, or, as shown here, creating special CDs with publicity results for them to view at home.

- Special focus stories in employee publications, such as newsletters, on your publicity efforts.
- A bulletin board in break rooms dedicated to rotating press clips generated in print media.
- Including special clips with paychecks.
- Creation and dissemination of special CDs with copies of publicity results.
- Playing audiotape interviews from radio shows in key employee areas.
- Sharing major publicity achievements during pre-shift employee meetings.
- Showing pictures of employees giving interviews to members of the media (especially being interviewed on television shows) in employee newsletters.
- If you know an air date beforehand for a television show showcasing one of your homes or a company representative, include a note to employees in paychecks encouraging them to watch the show.
- Send a broadcast mass e-mail to every employee on your company's e-mail system alerting them to upcoming special publicity results (such as television show air dates and magazines to look for stories in).
- If you have an exclusive feature story on your company in a magazine, have reprints made of the article and make them available to all employees.
- Post copies of press clips and upcoming schedules of air dates for shows featuring your homes on your company's website, or create a website exclusively for employees and include the success stories on the site.
- If your company has a "year in review" presentation for employees, make certain strong press results are included in the meeting presentation.

> Members of your sales team may be the most interested in receiving samples of press results. A good salesperson uses these success stories to share with customers and show how your company is getting solid press coverage, which should lead to more sales.

> "When we have a major media 'hit' we blanket our employees and customers with details. A great example is our involvement with the *Michael Holigan's Your New House* show. After donating windows for the construction of a custom home, we included the story in our employee newsletter, had special flyers sent to our dealers/distributors, and even handed out packaged popcorn with the show's air date at trade shows for potential customers.
>
> Just giving the windows away isn't enough. Maximizing our investment and promoting our involvement in a project really goes a long way towards creating a full-scale public relations campaign."
>
> Joy Frank-Collins
> Public Relations Manager
> Simonton Windows

Chapter 8

Promotions and Publicity

Sweepstakes, contests, coupons, incentives, rebates . . . the list is endless for the types of promotions that builders incorporate into their marketing activities each year. Whether you're offering manufacturer-based rebates on energy-efficient products or builder closeout specials, the opportunities for publicity are also unlimited.

One of the keys to maximizing your promotion and publicity capabilities is to put the power of public relations to work for you. If your company is launching a major sweepstakes, make certain to use press releases to generate publicity for the promotion. When the sweepstakes is finished, public relations can help garner awareness for the winners. Even if your company has launched a completely self-serving campaign, such as trying to sell more homes in a four-month period than ever before, there are story angles that can be incorporated into public relations campaigns to make the promotion appeal to the media and generate publicity for your company.

Ideas That Set Your Business Apart

Oftentimes with builders, the goal is to set your company and your homes apart from–and ahead of–competitors in the same product category. This can sometimes be achieved by hosting unique special events, gaining visibility for your company at community build projects, or even by being profiled in a trade publication.

Think about these ideas the next time you're searching for a unique promotion that can also generate publicity for your company:

- Can you work with your major kitchen appliance manufacturer to create a special cookbook? Have the manufacturer create the cookbook and then offer it as an incentive to people who come to view your gourmet kitchens in your model homes. Better yet, have local celebrity chefs host cooking classes and/or demonstrations in the kitchens of your model homes to increase visibility. Link them up with the local television station for a live shot and you have publicity opportunities for everyone involved!
- Look toward the future and help train tomorrow's builders. Contribute unwanted product and guest speakers to vocational and trade school programs in your area. There's nothing better than getting tomorrow's workforce familiar with your company today!
- Sponsoring competitions and scholarship programs is also a terrific way to convey a message that your company cares about quality in the housing

Figure 8-1. Are you installing high-end kitchen appliances in your home? If so, contact the manufacturer. See if they can work with you on creating a cooking demonstration in your market, exhibit at a local home show or even create a cookbook for your customers. "Push the envelope" to set your business apart in the marketplace.

industry and in your community. Consider hosting a vocational technical class at your jobsite once a quarter for hands-on experience. Organize and host a school scholarship program for future builders. Start a summer internship program for students to get jobsite experience.

- Host a contest or competition on your jobsite for the general public or other builders. Involve a charity and you've got instant headlines. For years the Industry Education Alliance has sponsored nailing contests at trade shows that create awareness and interest in the sponsoring company's products. You can do the same at your jobsites.
- Looking to capture the interest of people in a specific market? Research home improvement radio shows that cover the market. Then offer your company spokesperson as a guest for a question-and-answer session on trends, products, and building practices in your area.

> Want to gain repeat traffic on your website? Then make it work for you. Create promotions and giveaways that encourage your target audiences to keep coming back. Sponsor question-and-answer chats with leading experts. Update your site with a celebrity corner and make celebrities out of your customers by running their stories about owning your homes. The ideas are endless—and many publicity and promotion ideas on your site could end up as great stories for the media.

> Take the lead from Hearthstone Advisors, or take advantage of their generosity. In 1999, Hearthstone Advisors, one of the nation's largest home building project financiers, partnered with *Builder* magazine to create an annual awards program. The joint program rewards builders who give back to their communities and serves as encouragement for other builders to do the same.
>
> The Hearthstone Builder Lifetime Public Service Award program recognizes individuals who "give their time, money, and resources to make their communities better places to live." Combined sponsors of the award—including several major building product manufacturers—have pledged to give more than $1 million over five years to the winning builders' favorite charities. According to an article in *Builder* magazine, this makes the program the largest philanthropic cash prize in the home building industry.
>
> Certainly Hearthstone Builder set themselves apart by creating this award. But they also opened the door for home builders nationwide to receive recognition. Nominations are accepted each March. A listing of criteria, a nomination form, and more information can be obtained by visiting www.builderonline.com.

- Don't laugh, but sponsoring youth-oriented sports teams in your community is a good way to build relationships with the parents of the kids involved. Work the promotion to help your business. Don't just put your names on the T-shirts and forget about the team. Host a special end-of-season party for the parents and team members at your clubhouse or model home, offer buyer incentives to parents of team members, attend the games with literature on your homes. Really put effort into sponsorships and they'll pay back.

Rule #1: Maximize Exposure

Whether your company participates in a community project, sales-building promotion, or publicity event, always focus on rule #1: Maximize your exposure. Companies should push the boundaries when involved in marketing and public relations programs. Grab any and every opportunity to get your company's name in the marketplace.

Need an example? Let's create a fictitious company called Robeson Builders. The company primarily builds custom homes for high-end homeowners.

The headquarters for Robeson Builders is located in Eugenia, Kentucky—population 125,300. Each year Robeson Builders is asked to sponsor a home in the local home builder association's parade of homes. Because they're good community citizens, they agree and tackle the project with gusto. In return, they get a sign acknowledging their participation plus all the usual promotional elements associated with a parade of homes.

This year Robeson Builders has hired a professional public relations person as part of the company's marketing team. When it comes time to participate in the annual parade of homes, the public relations person tackles the project with enthusiasm. She arranges for the following:

- Interviews with the builder and architect involved in the home for the local newspaper and television station. She focuses on unique design elements of the home, construction quality, and 21st century building products used in

the home. She turns this into a press release for the local Eugenia media and sends it out before the home is open for public tours.
- Once the home is complete, the public relations person arranges professional photography of the project. She includes the story, with photos (from the press release), on the website. She also sends the release and photos to the key manufacturers that Robeson Builders has been promoting–and several of the manufacturers include the story on their websites.
- The public relations person sends the story and images to the Kentucky state builder magazine and to *Custom Home* magazine to seek coverage.
- At the homesite, in addition to the signage, literature is made available to visitors during the parade of homes. A representative from Robeson Builders is on site to answer questions during all open show home hours.
- Robeson Builders wants to connect with people who tour the home, so they hang framed pictures throughout the home of the construction phases and the building team. They also use table-top signage to offer tips and suggestions for decorating and upgrades to the home, including the cost to the homeowner.
- To generate more attention during the parade of homes, the company issues a press release to local and regional media on different elements of the home, such as what they did to add curb appeal and enhance the resale value of the home.
- Although there has never been an awards program in the past connected with the parade of homes, Robeson Builders has convinced the local home builder association to initiate one. They then enter in several categories.
- The Eugenia Parade of Homes is sponsored by a local radio station. So the public relations person makes arrangements for an interview with the builder on the radio station to help pre-promote the tours of the homes. Listeners are encouraged to call in with their questions about everything from home maintenance to selection of building products. A follow-up broadcast is scheduled live from the show home on the opening day of the parade.
- The public relations person designs postcards with a photo

> When Morris Builders in Rockford, Michigan, won an award from the Better Business Bureau of western Michigan, they maximized their exposure by informing the trade magazine *Remodeling*. Subsequently, the editors ran a half page story on Morris Builders. Within the story, Joy Morris mentioned that the local newspaper and television publicity they received was extensive.
>
> "It turned us into celebrities," said Morris. "Everywhere you'd go, everybody had seen it." To further maximize the award, the *Remodeling* story reported that Morris Builders sends out "Best In Customer Service" labels on all their correspondence. To Kirk and Joy Morris, one of the best aspects is that they attribute the award to seven leads for new business, plus an $80,000 job from a past customer!

> If you don't have them already, make sure to get a variety of labels and stickers with your company logo. These can be used on mailings, press kits, tickets, and other materials distributed by your company during the year.

> Many other savvy builders throughout the country are just as eager to gain positive publicity as you are. Read through trade publications and you'll be able to quickly identify some industry leaders who have "made a name for themselves" by being available to the media. Those are the kind of open-minded professionals your company needs to link up with.

> To maximize your company's exposure, be sure to send copies of nationally placed media stories involving your company to your local media. They may pick up on the story and run another story on your role as a national spokesperson within the industry.

of the home or an artist's rendering. Robeson Builders then sends the postcards to past customers and potential customers, including all realtors in the area. Going a step further, the postcards (which promote visits to the show home) are sent to their local vendors for use as counter cards and bag stuffers to support attendance at the show home.

Think that's all that can be done to promote the involvement of Robeson Builders in this parade of homes project? Think again.

Before construction started on the home, the public relations person made arrangements for the local technical school class to visit the construction site and learn several aspects of building construction, and a local television crew was asked to cover their visit. The public relations person next arranged for the local Master Gardener's Club to handle the landscaping outside as a class project. After that, Robeson Builders hosted a ribbon cutting and grand opening ceremony to launch the opening of the show home.

As you can see, the ideas are endless for maximizing exposure and publicity opportunities. Simply picking one or two of these ideas would generate added media and public attention for Robeson Builders. Putting the full attention of a public relations professional on the project guarantees additional company awareness and return on investment for participating in the parade of homes.

Not every project is as comprehensive as the one described for Robeson Builders. Remembering to keep your company and product names in the forefront of your marketing and public relations efforts at all times will help your company maximize exposure. Whether it's having employees wear your company logo T-shirts on the jobsite or keeping your logo prominently displayed on letterhead, press kits, and slide mounts, maximizing exposure of your company can lead to more awareness and greater publicity.

Chapter 9

Planning for a Crisis

Here's the moment of truth. Does your company have a written crisis plan document? If not, you fall into the "ostrich category."

You're not alone. Many builders believe a serious emergency will "never happen to us" or, worse yet, "we've never had a problem before, so we're safe." This is worse than just sticking your head in the sand; it's the equivalent of being an ostrich wearing rose-colored glasses.

As events have shown us in recent years, there is no safe harbor from crisis situations. Churches, schools, office buildings, and other previously thought of "safe havens" have come face-to-face with very serious crisis situations. Merely saying the words "Columbine," "September 11," and "anthrax" immediately trigger memories of terrible crisis situations.

Still thinking that a crisis situation is a remote possibility for your company? Ask any builder who has dealt with the EIFS or a mold/mildew situation how serious it is. Ask a builder who has had a fire at a construction site or the death of an employee as a result of a construction accident how serious it is. A crisis situation can impact a builder in numerous ways. The key is being ready to handle the negative situation so it doesn't spin out of control.

Having a well-practiced crisis plan in place allows you and your management team to have some level of control over the results of a crisis. In the face of tragedy, a crisis plan acts as a security blanket of knowledge, giving you the confidence to work through emotions, confusion, and an extraordinarily difficult situation.

It CAN Happen to You: A Look at Scenarios Facing Home Builders

Still not convinced a crisis can impact your company? Consider these scenarios:

- No building site is exempt from the possibility that a fire will occur at any specific time (day or night) or that a disgruntled employee will orchestrate one. What if a home has faulty electrical wiring installed by a subcontractor and burns to the ground? How will you deal with this situation and other homes wired by the same subcontractor?
- Many builders have superior safety records, but no company is completely accident-free. How would you deal with the media when explaining that a forklift operator, who tested positive for drugs afterwards, ran over and killed an employee while unloading a lumber shipment?

- EIFS. Mold. Mildew. These are all current concerns for builders. The media have made these into big stories nationwide. What will you do if a reporter visits your office and starts asking you questions about your installation practices related to these issues? What if an unhappy homeowner calls the media to report that the home you constructed is "sick" with mildew and mold?
- Mother Nature does not cut builders any slack. Tornadoes can rip roofs off homes under construction. Hail can damage products stored outdoors. Hurricanes can devastate a development. Heavy flooding can damage facilities, prevent employees from getting to work, or trap them at jobsites.
- Imagine that the driver of a delivery truck from your local dealer accidentally "guns it" while the truck is in reverse, resulting in the truck doing massive damage to the home under construction and injuring several of your employees.
- Two words that make the skin grow clammy for any builder or building products manufacturer are "product recall." What happens if disgruntled employees at one of your suppliers tampers with products during the manufacturing process, causing failure of the products? Are you prepared to track down products installed and replace them? How about handling the media in such a situation?
- What if your product was sabotaged? Everyone remembers the famous tampering incident with Tylenol back in the 1980s. Think it can't happen in the building industry? Think again. What if one of your suppliers has a bad batch of products that are a key component for your homes, such as window hardware or particleboard resin. You find out that these products were intentionally damaged by employees frustrated that the company would not become unionized. The situation should not impact your company but it does. Because now you have inferior components in the windows you installed. Your company's sales, your customers' safety, your reputation, and future are at stake.
- In these economically stressful times, many companies have been forced to lay off employees–sometimes only temporarily. How a company handles this situation internally and externally positions them for future success. Are you prepared to deal with disgruntled employees who picket against your homes outside one of your developments or show homes?
- Many different people work on jobsites– your employees, subcontractors, etc. What

> "Crisis planning is something all builders and companies should do. It doesn't matter whether you're a huge, multi-million dollar company or a one-person builder. If you're in business, you need to have a practiced plan for dealing with urgent situations."
>
> Mary Balice
> Director of Training
> Citigate Communications

happens if a personal feud escalates and a fistfight (or worse) results in people getting hospitalized? When the television camera crews arrive to interview you, what will you say?
- Want another word that has made builders cringe over the years? Asbestos. What if you and your company get sued because homes you constructed 25 or more years ago contained asbestos that was making homeowners sick?

You've Got to Have a Plan

There's simply no way around it: Every builder needs to have written emergency and crisis plans. Emergency plans are operational documents that are posted and shared with employees. They cover such things as evacuation procedures of a facility during an emergency, safety policies supported and enforced by your company, and action steps taken at any of your facilities or on the jobsite to assure the security and safety of all employees.

Crisis plans are generally created for upper level management team members to help them handle the aftermath of an emergency. How a company conducts itself during and after a crisis greatly determines how the company will rebound from the situation.

Preparing for the Unexpected

Let's look outside our industry for examples of properly and improperly handled crisis situations. Who are the losers? Exxon. After the crash of the Exxon Valdez oil tanker, the company's refusal to immediately send senior management to Alaska was viewed as callous. It appeared that Exxon displayed a lack of concern and responsibility. As a result, thousands of Exxon customers cut up their Exxon credit cards, returned them to the company, and vowed never to use the company's product again.

Remember several years ago when an 81-year-old woman spilled a cup of scalding coffee on herself at McDonald's and suffered third-degree burns? She won a multi-million dollar lawsuit against the fast food company. However, McDonald's lost more than money. Their positive reputation was tarnished when they were portrayed as being insensitive and uncaring.

There are dozens of other examples: The poor handling by Denny's Restaurants of racial discrimination claims by African American customers. Dow-Corning's reluctance to disclose information on the potential hazards of their silicone breast implants. Accusations and investigations into the proper distribution of funds raised by the American Red Cross in the aftermath of September 11, 2001 terrorist activities.

Now consider some "winners." The September 11th events bring to mind a real winner in the handling of crisis situations. Described afterwards as "America's Mayor," Rudy Giuliani stepped in to calm his city in the wake of horrendous terrorist activities. His compassionate, straightforward, and responsible stance helped give the city–and the country–confidence and calm fears.

> Don't wait another day. Decide now who will be your company's spokesperson if the unexpected and unthinkable happens.

After the attack on New York City, Mayor Giuliani was named "Person of the Year" by *Time* magazine for his handling of the crisis. In the cover story, *Time* reported that Giuliani had planned for different disaster scenarios throughout his entire mayoral term. Practicing for the unthinkable helped focus Giuliani in the immediate hours after the attacks. As the *Time* story puts it, "Giuliani treated the public like grownups, offering unvarnished information and never having to backtrack. When he told people not to panic, they didn't."

Looking for another winner? Consider Pepsi-Cola. In 1993 a nationwide hoax by a man in Tacoma, Washington–who asserted that he had found a syringe inside a can of Diet Pepsi–pushed the Pepsi crisis team into action. The management team at PepsiCo cooperated fully with investigators from the U.S. Food and Drug Administration, gave numerous interviews, and widely distributed video news releases to media nationwide showing the fast-paced assembly of Pepsi products. The videotape alone showed how impossible it would be for their product to be tampered with.

Imagine this scenario: The president of a major airline is donating his day to working on a local Habitat For Humanity project in Atlanta, Georgia. Management team members are with him when he gets the call that a plane has crashed in the Florida Everglades. That's what happened to Lewis Jordan of ValuJet Airlines in 1996. Although the accident was a tragedy, the company handled the situation in a compassionate and responsible manner. The president was immediately available to the media and families of victims. ValuJet Airlines promised to cooperate fully in the investigation and Jordan personally flew to the crash site. Although the company ultimately had difficulty rebounding from this tragedy, their responsible and caring handling of the situation positioned them favorably at the time of the crisis.

What commonalties do the winners have? Plans. Each company and city planned–and practiced–for the day a crisis would strike.

Sharing the Plan With Others

Creating a comprehensive crisis plan can be a task your internal management team handles as a group. It can also fall to your marketing or public relations departments. There is also an abundance of professional crisis planning companies nationwide that can come in, analyze your company's business needs, and create a plan for you.

A solid working crisis plan incorporates the following elements:

- Outlines the key members of the crisis communications team and their roles during a crisis.
- Identifies and analyzes potential crisis situations for your company and categorizes them.

- Includes a listing of resources available to assist your company in the case of a crisis, such as:
  - Contact information for local police, fire, and emergency authorities
  - Churches within five miles of your headquarters
  - Hospitals within 20 miles of your jobsites
  - Grief counselor resources within 30 miles of your headquarters
- Includes a selection of forms and checklists available to assist your company employees in case of a crisis, such as:
  - Situation report form
  - Bomb scare checklist form
  - Forms for recording general public and media inquiry calls
  - Background history press release on your company
  - Updated company fact sheet

> "Getting a plan in writing is critical. The plan should be shared with key employees. It can definitely serve as a safety blanket during the emotional times of a crisis. Practice for the worst and you're setting your company up to successfully handle any negative situation."
>
> Jeff Braun
> Vice President
> The Ammerman Experience

Once you have a written plan in place, it's important to make sure key company team members have copies of the plan and keep it with them at all times. Some companies have their plans on CDs, disks, or on Palm. When you're vacationing in Aruba and get the call that there's a crisis back home, you want the plan with you. Remember the saying shared earlier in this book, "Plan the work and work the plan"? The same holds especially true with a crisis plan.

Designating a Spokesperson

During a crisis situation the credibility of your company may be at stake. It's extremely important that the top-ranking person at your company serve as the spokesperson during a crisis. If not, people will wonder where that top-ranking person is, whether he or she is "hiding" and why he or she isn't speaking with the press.

For everyday media spokesperson activities such as to talk about a new development being planned or the expansion of your business, it may be fine to designate a senior management member to

> Create your crisis plan according to positions, not individual people. This way, when people leave the company or change jobs within the company, the plan does not need to be updated, just the individuals with new responsibilities.

> Like the day when President Kennedy was shot, people will always remember where they were on September 11, 2001. For the management team of Style Solutions Inc., an Ohio-based millwork manufacturer, they'll recall how surreal the visions on the TV screen seemed, how they thought it was just another part of the day-long crisis planning workshop they were participating in.
>
> "We came out of the meeting room to take a break and saw people gathered around the hotel lobby television," remembers Arlan Yoder, former president of Style Solutions. "I thought the crisis training team was clever to incorporate such a dramatic twist to our session."
>
> Moments later, as reality sunk in, the Style Solutions team cancelled the workshop and headed back to their offices to check on traveling salespeople and employees. "How bizarre is it that while we were training to handle the impact of a crisis at our company, the world's largest crisis was taking place," wonders Yoder.
>
> Gathering again in late November for the rescheduled training session, Yoder and his managers had a new respect for the trainers and time invested in the session. "Suddenly, with what happened on September 11th it didn't seem so implausible that we should be training to handle situations like transportation shutdowns due to terrorist activities," says Yoder. "Originally we were practicing emergency situation scenarios dealing with facility fires and storm-related damage. Now we've expanded our plans to include potential biohazard situations and loss of labor due to military reserve call-ups."

handle media interviews. However, if you ever find yourself facing media questions related to a crisis situation, the CEO, president, or owner should be handling the media inquiries and interviews.

Because a crisis situation is so serious in nature, this is not the time to practice at being a spokesperson. If you have a large company and have determined that the top management team members will only speak to the media in case of a crisis, make sure to prepare, train, and rehearse them for that potential on a regular basis. Bring in an outside trainer with expertise in crisis situations. You want your entire team to feel comfortable handling media inquiries, not be caught with a "deer caught in the headlights" look.

Just because the top person at your company should take responsibility and answer media questions directly during a crisis doesn't mean he or she has to stand alone. If you're involved in a serious crisis situation that requires a press conference or multiple media interviews, it's perfectly acceptable for the CEO or president of the company to allow one or two other company leaders to participate in the media interviews. For example, someone in charge of structural engineering at your company or your vice president of construction may be a valuable addition to assist the CEO if there's a crisis wherein the structural integrity of your homes comes under fire. It is key that the CEO present a united front with his team and use their expertise to help diffuse and explain the situation.

Practice, Practice, Practice

Did you ever learn to ice skate as a kid? If you tried to skate again today, after years off the ice, what do you think would happen? Most people readily admit they'd fall down at least once. Do you think you could play Mozart on the piano today because you took lessons on Friday afternoons as a child? Probably not.

The same theory exists for a crisis plan. If you don't practice, you're going to fall down, and probably strike some sour notes along the way. Companies that create a crisis plan and then bury it in a desk drawer are doing themselves a disservice. For a company to be "crisis ready" the crisis plan needs to be a living, breathing document that is upgraded and changed throughout the year. The plan also must be shared with new crisis team members as they join the company.

One of the easiest ways to keep your management team sharp on handling a crisis is to start out on the right foot when the plan is created. Have a full-day session dedicated to reviewing all aspects of the plan, learning spokesperson training skills, and role-playing. Once this initial meeting has taken place, establish quarterly dates for two-hour update sessions. The review meetings can be led by your public relations staff person or an outside consultant. Each session should put forth a different crisis scenario and then challenge the team to solve it. When you're at a practice session, make sure everyone sits in the "hot seat." Develop scenarios that requires your team to think hard through unique problems, answer media questions, and come up with action plans. Create different scripts that take your lead spokesperson out of the country on travel so that secondary spokespeople also get training on handling media inquiries.

Looking for some ideas to practice at your company? Consider adapting any of the following scenarios and then determine how well your team reacts:

- A tornado rips off the roof of your company headquarters. A

> "Companies that put a CEO or president in front of the media during a crisis situation who normally doesn't handle media interviews may find themselves facing two crises. If a company leader flubs a media interview, looks flustered or out of control during a television interview, or balks at handling intense media questions, the entire credibility of your company may suffer. From there it could become a downward spiral. Making sure company leaders are prepared to handle media interviews should be a key goal of any builder company."
>
> Betty Christy
> Former Vice President of Public Affairs
> National Association of Home Builders

TIP: Local NAHB chapters throughout the nation routinely conduct spokesperson training sessions to prepare builders for handling themselves in front of the media—both during crisis situations and everyday interviews. Find out when your NAHB chapter will conduct a session and make it a priority for several members of your management team to participate.

> "The NAHB spokesperson/crisis training program taught us how to be on the offense with the media and not the defense. We were put in confrontational situations with cameras and reporters. It was great practice that showed us how to prepare our messages before giving interviews, settle down and get our thoughts in order, and present our best image to the media.
>
> I would recommend this training program to any builder. We mandate that the president and vice president of our local home builders association take the course. Sometimes builders are vilified in the media on certain issues. This has helped everyone I know prepare a message, stay on their key topic points, and present a strong appearance to the public."
>
> Charles Clayton
> President
> Charles Clayton Construction, Inc.

dozen employees are injured during the storm and at least half the building has been heavily damaged. How are you going to handle media inquiries about the status of your employees and if your company will stay in business?

- As the builder of multi-family homes, you learn that the oriented strand board (OSB) you purchased and installed has broken down on the jobsite. Pieces are literally flaking off. A day later you find out there's a national recall by the manufacturer. How are you going to handle going back in and tearing out all the OSB in the multi-family homes you constructed? What will happen to the families? And can you handle the national press attention?
- During a holiday weekend when your business is closed, a fire breaks out on the jobsite of a multi-million dollar custom home. Damage is minimal because it's caught in time. However, local fire officials determine this was an act of arson by a team of disgruntled subcontractors. How are you going to handle the ensuing media inquiries?
- During a sting operation, it turns out that the CEO for your publicly held company accepted cash bribes from a supplier to use their products. The news is everywhere and the media are requesting a press conference to find out how this happened at your company and what will happen to the CEO. Are you ready?
- You're trying to get the land in a local community to build a development. Citizens are outraged because they don't want another development nearby. They're picketing your existing developments and meeting with the press to express their concerns. How will you handle this situation?
- The city licensing inspector your company has used for a decade has just been arrested for reportedly taking bribes. Although this has not occurred at your company, you appear guilty by association in the press. Interest in home sales has immediately dropped off, plus existing homeowners are nervous that their homes might not be safe. Are you ready to meet with the press?
- Even after years of sensitivity training and counseling, one of your lead supervisors has made disturbing racial remarks to other employees. Fed up

with the treatment, the employees have gone to the local newspapers to complain. How will you handle the ensuing media allegations that your company is made up of bigots and racists?

> Make it a priority from the executive level down to schedule two half-day practice crisis sessions each year. This will keep everyone focused on the plan and provide new players with the opportunity to interact with the group during mock situations.

Remember, don't be the company that resembles the ostrich wearing rose-colored glasses. These scenarios–and many more–can definitely happen to any builder. No one is exempt. However, with a proper, well-practiced crisis plan in place, your company can get through tough times.

Do's and Don'ts of Media Interviews During a Crisis Situation

Whether it's an extremely personal situation, such as a company plane crashing and killing several key management team members, or a less emotional predicament, such as a work stoppage, a crisis situation gets the adrenaline pumping. During the initial stages of the crisis situation, when people are trying to comprehend the magnitude of the problem, emotions run high.

Oftentimes, during these first critical hours the most devastating mistakes are made by management team members. Distancing the company from the situation and saying "no comment" are two of the most dangerous reactions a company can have to a crisis. This is the time for builders to put a human face on their company, to express their compassion and concern over the crisis situation.

In a majority of cases, a crisis situation will require a company spokesperson to speak directly with one or more media representatives. Media interviews related to the crisis can be handled one-on-one in person, over the phone, or as a group in a press conference setting. Regardless how large or how small the crisis situation is, there are a number of "tips" for handling media interviews during a difficult situation.

Citigate Communications in Chicago, a firm dedicated to assisting companies handle crisis situations and spokesperson training, offers these ideas:

Before speaking with a media representative, ask yourself the following questions:

- What are my goals for this interview?
- Who is my audience?
- What are the three most important points I want this audience to hear?
- What personal examples do I have to support these points?
- What questions is the reporter most likely to ask me?

Answering these questions for yourself–and as a crisis team–before the media interview will set you up for success.

> Never make a comment to anyone—especially the media—like, "We've got insurance to handle bad situations like this." This kind of statement makes your company appear cold, callous, and uncaring about the human emotions and difficulties involved in a crisis situation. Worded carefully however, you could say that "Our hearts go out to everyone involved in this difficult situation. Our management team is doing everything possible to assist the people affected. Fortunately, we are a responsible company. That means we have insurance coverage available to assist the victims of this tragic situation."

Citgate Communications recommends considering the following checklist when a reporter calls you:

Do:
- Tell the reporter you'll call back in 15 minutes and use the time to get prepared for his questions.
- Return the reporter's call within the timeframe you promised.
- Be quotable for the reporter.
- Use the interview as an opportunity to meet your agenda and put forth your important points, not just answer the reporter's questions.
- Speak clearly and plainly. Eliminate industry jargon that the common person will not understand.
- Be honest and open with the reporter. If you don't know the answer to something, tell the reporter.
- Stay in control of the interview.
- Ask the reporter to call you back to confirm your quoted comments.

Do Not:
- Ask the reporter for a list of questions before the interview.
- Say anything you don't want used in the story. There's no such thing as "off the record."
- Ask to see the story before it is published.
- Panic. Relax, reinforce your messages, and stay in control.
- Assume the reporter understands the building industry or its issues. Educate the reporter and make him clearly understand the issues.
- Talk in code or jargon. Instead, explain ideas, issues, and terms in simple language.
- Ramble. Keep your answers short, simple, and to the point.
- Answer a negative or hostile question defensively. Even if a reporter baits you, remain calm and steer the interview back to your key points.
- Promise to get back to a reporter with additional information unless you actually plan to do so.
- Play favorites. You should not give information to one reporter that you are not making available to all reporters.

Following these guidelines sets you up for success when serving as a media spokesperson. To sum it up, a spokesperson during a crisis situation should be prepared to answer media inquiries in a compassionate, honest manner. Stay in control and offer the media quotable statements that clearly support the position of your company.

Appendix

Resources

Media Lists
Bacon's Information, Inc.–800-PR-MEDIA or *www.bacons.com*
eNEWSRELEASE.com–888-607-9101 or *www.enewsrelease.com*
INKDB.com–866-346-3246 or *www.inkdb.com*
Media Distribution Services–800-MDS-3282 or *www.mdsconnect.com*
MediaMap–617-393-3200 or *www.mediamap.com*
PIMS–866-GET-PIMS or *www.pinsinc.com*
Press Access, A LexisNexis Company–800-227-4908 or *www.pressaccess.com*
U.S. Newswire–800-544-8995 or *www.usnewswire.com*

Clipping Services
Allen's Press Clipping Bureau–415-392-2353
Bacon's Information, Inc.–800-PR-MEDIA or *www.bacons.com*
Burrelle's Information Services–800-631-1160 or *www.burrelles.com*
CompetiveEdge–860-726-1047 or *www.clipresearch.com*
INKDB.com–866-346-3246 or *www.inkdb.com*
Luce Online, Inc.–800-518-0088 or *www.luceonline.com*
Luce Press Clippings–800-528-8226 or *www.lucepress.com*
Multivision, Inc.–800-560-0111 or *www.multivisioninc.com*

Crisis Communications and Spokesperson Training Services
The Ammerman Experience–800-866-2026 or *www.ammermanexperience.com*
Christy Consulting–703-492-6562 or *www.e.christy.verizon.net*
Citigate Communications–312-944-7398
The Communications Center–800-929-SPPI or *www.susanpeterson.com*
High Impact Television–949-852-5977 or *www.HighImpact.Tv*
The Newman Group–212-838-8371 or *www.newmangroup.com*
Reputation Management Associates–614-486-5000 or *www.media-relations.com*

Manuals and Books
The Associated Press Stylebook and Libel Manual–Available from The Associated Press, 50 Rockefeller Plaza, New York, NY 10020

Communications Industry Associations

Public Relations Society of America (PRSA)–212-995-2230 or www.prsa.org
Public Relations Student Society of America (PRSSA)–212-995-2230 or www.prssa.org
International Association of Business Communicators (IABC)–415-544-4700 or www.iabc.com
The Association for Women in Communications, Inc. (AWC)–410-544-7442 or www.womcom.org
National Association of Real Estate Editors (NAREE)–561-391-3599 or www.naree.org

Index

A

Advertising
  direct mail, 6
  marketing uses of, 3–4
  public relations and, 4, 6, 58
  story placement and, 58
  strategic, 4, 6–7
Agencies, *See* Public relations agencies
Asbestos, 119

B

Brochure, 25
Budgeting, 15, 35–36
Builder
  community image of, 84–85
  crisis that face, 117–119
  expert status of, 76–78
  joint efforts with other builders, 92–94
  literature of, 34–35
Builder associations
  contacting of, 128
  websites for, 91–95
  working with, 90–92
Building product manufacturers
  joint efforts with, 94–96
  product donation by, 8–9
  public relations expectations of, 8–9
By-lined stories, 27–28

C

CD-ROMs, 27, 31
CEO, 65, 122
Charitable giving
  considerations for, 80–83
  success of, 88
Clipping services, 9–10, 37, 127
Clips
  analysis of, 10
  sales use of, 38
  uses for, 38–39
Communication with employees, 109–110
Community relations
  benefits of, 79
  company image, 84–85
  description of, 79
  employee involvement in, 87–88
  local opportunities for, 82–84
  maximizing of, 80, 87–88
  projects for, 80–81
  reasons for, 79–80, 84–85
  sponsoring of sports team, 114
  success of, 88–89
Company
  community image of, 84–85
  expert status of, 76–78
  literature of, 34–35
Consultants, 16–17
Crisis
  description of, 117
  media interview guidelines during, 125–126
  resources for, 127
  types of, 117–119, 123–125
Crisis plan
  elements of, 120–121
  importance of, 119
  practicing of, 122–125
  preparing for the unexpected, 119–120
  resources for, 127
  scenarios for, 123–125
  sharing with others, 120–121

Crisis plan (*continued*)
  spokesperson, 121–122, 126
  successful examples of, 119–120

D
Dealers, 96–99
Deskside briefings, 31–32
Direct mail, 6
Donation of products, 8–10, 81

E
Editors, 55–58
Emergency plan, 119
Employees
  communication methods, 109–110
  community relations by, 87–88
  company vision shared with, 106–107
  lay-offs, 118
  promotion by, 107–108
  publicity results shared with, 110–111
  tools for reaching, 108–110
Exhibit booths, 86–87

F
Fact sheet, 24
Former customers, 102–105

H
Home builder
  community image of, 84–85
  crisis that face, 117–119
  expert status of, 76–78
  joint efforts with other builders, 92–94
  literature of, 34–35
Home builder associations
  contacting of, 128
  websites for, 91–95
  working with, 90–92
Home improvement radio shows, 113
Home shows, 85–87
Home tours, 33

I
Industry trends stories, 29–30
Interns, 13–14
Interviews
  during crisis, 125–126
  radio show, 87
  television shows, 68–69
Invitations, 42–43

J
Job site
  accidents at, 117
  fighting at, 118–119

L
Lay-offs, 118
Lead times, 56
Literature, 34–35

M
Magazines
  by-lined stories, 27–28
  consumer, 51–52
  editors, 56–57
  exclusivity of stories, 75
  special projects with, 99–102
  time requirements for, 101
  topics not covered by, 74
  trade, 55–56
  types of, 50–52
  unique projects shared with, 75–76
  working with, 99–102
Manufacturers, *See* Product manufacturers
Marketing
  advertising, 3–4
  public relations' role in, 2–7
  tools for, 2
Media, *See also* Newspapers; Radio shows; Television show
  CEO access, 65
  current trends, 72–74
  focus of, 59

follow-up with, 62
home tours offered to, 33
interviews during crisis, 125–126
magazines, 50–52
story evaluation by, 47–48
television shows, 52–53
types of, 50–57
understanding of, 47
Media binders, 30–31
Media directories, 54
Media lists, 62, 127
Media members
answering questions of, 60–61, 68
attention-seeking plan for, 63–65
contacts, 62–64
dealing with, 58–61
expectations of, 59–61
guidelines for working with, 67–70
pet peeves of, 69–70
requests of, 60, 64–65
responses to, 70
unique projects shared with, 75–76
Media placement services, 34
Media relations
elements of, 47–50
honesty in, 58–59
lack of guarantees, 57–58
spokesperson, *See* Spokesperson
strategies for, 61–62
Mold, 118

N
News release, 23–24
Newspapers
contacts at, 62–63
home improvement columns in, 53–54
publicity opportunities, 3
stories in, 53–54
Newsworthiness of stories, 70–72, 74

O
Outsourcing, 14–16

P
Partnering
with building product manufacturers, 94–96
with dealers, 96–99
with former customers, 102–105
with home builder associations, 90–92
with other builders, 92–94
Past customers, 102–105
Photographs, 24–27, 102
Pitch letter, 64
Plan
crisis, *See* Crisis plan
public relations, 19–21, 76–77
Press clips, 9
Press kit, 24–25
Press releases
community relations use, 87–88
conciseness of, 64
customizing of, 61
description of, 21–24
lead time for, 56
writing of, 64
Product
donation of, 8–10, 81
recall of, 118
Product manufacturers
joint efforts with, 94–96
product donation by, 8–9
public relations expectations of, 8–9
Project homes, 52
Promotions
description of, 112
ideas for, 112–114
maximizing of exposure, 114–116
Public relations
advertising and, 4, 6, 58
affordability of, 4
budgeting for, 15, 35–36
builders who should use, 7–8
checklist for, 6–7
clipping services for evaluating, 37
definition of, 1–2

Public relations (*continued*)
  expectations for, 8–10
  goal setting for, 17–18
  interns, 13–14
  long-term, 8
  marketing uses of, 2–7
  on-staff communications/marketing people for, 12–13
  outsourcing of, 14–16
  plan for, 19–21, 76–77
  prioritizing of, 17–18
  purpose of, 1
  success measurements, 37
  target audience for, 18–19
Public relations agencies
  locating of, 15–16
  organizations, 15
  outsourcing to, 14–16
Public relations consultants, 16–17
Public relations people
  communications/marketing persons, 12–13
  creativity of, 76
  deskside briefings, 32
  interns, 13–14
  on-staff, 12
  professional, 11–12
  recommendations for, 11–12
  selection of, 65–66
  traits of, 11–12
Public relations practitioner, 1–2
Public Relations Society of America, 2, 15, 34, 128
Public Relations Student Society of America, 14, 128
Public relations tools
  by-lined stories, 27–28
  CD-ROMs, 27, 31
  company literature, 34–35
  deskside briefings, 31–32
  home tours, 33
  industry trends stories, 29–30
  media binders, 30–31
  media placement services, 34
  photographs, 25–27
  slides, 25–27
  transparencies, 25–27
  videos, 25–27
  website, 27
Publicity
  current trends, 72–73
  maximizing of exposure, 114–116
  newsworthiness of story, 70–72, 74
  opportunities for, 77–78

R

Radio shows
  home improvement, 113
  media contacts, 63
  public relations opportunities, 53
Recall of product, 118
Releases
  news, 23–24
  press, 21–24
Resources, 127–128

S

Shows
  home, 85–87
  radio, *See* Radio show
  television, *See* Television show
  trade, 87
Slides, 25–27
Special events
  defining of, 40
  invitations for, 42–43
  logistics of, 40–44
  maximizing of, 46
  tips for planning, 42, 44
  types of, 45–46
Spokesperson
  assigning of, 65–67, 121–122
  crisis plan, 121–122, 126
  radio show interviews, 87
  selection of, 65–67
  strengths of, 76
  training of, 67, 127
Sponsoring of sports team, 114

Stories
   advertising effects on, 58
   by-lined, 27–28
   exclusivity of, 75
   industry trends, 29–30
   media evaluation of, 47–50
   newspaper, 53–54
   newsworthiness of, 70–72, 74
Strategic advertising, 4, 6–7

T

Target audience, 18–19
Television show
   industry experts, 77
   interview guidelines, 68–69
   opportunities with, 52–53
   sharing success of, 39
   special projects with, 99–102
   time requirements for, 101
   types of, 53
   working with, 99–102

Trade magazines, 55–56
Trade professionals, 55
Trade shows, 87
Transparencies, 25–27
Trends
   description of, 73–74
   stories regarding, 29–30

U

Unique projects, 75–76

V

Volunteerism
   considerations for, 80–82
   success of, 88

W

Websites
   company promotion using, 27
   home builder associations, 91–95